# Lecture Notes in Mathematics

Edited by A. Dold and B. Eckmann

1087

Władysław Narkiewicz



# Uniform Distribution of Sequences of Integers in Residue Classes

Springer-Verlag
Berlin Heidelberg New York Tokyo 1984

**Author**

Władysław Narkiewicz
Wrocław University, Department of Mathematics
Plac Grunwaldzki 2–4, 50-384 Wrocław, Poland

AMS Subject Classification (1980): 10A35, 10D23, 10H20, 10H25, 10L20, 10M05

ISBN 3-540-13872-2 Springer-Verlag Berlin Heidelberg New York Tokyo
ISBN 0-387-13872-2 Springer-Verlag New York Heidelberg Berlin Tokyo



Printing and binding: Beltz Offsetdruck, Hemsbach / Bergstr.
2146/3140-543210

To my teacher

Professor Stanisław Hartman

on his seventieth anniversary

## INTRODUCTION

The aim of these notes, which form an extended version of lectures given by the author at various places, is to present a survey of what is known about uniform distribution of sequences of integers in residue classes. Such sequences were studied since the beginning of this century, when L.E.Dickson in his Ph.D. thesis made a thorough study of permutational polynomials, i.e. polynomials inducing a permutation of residue classes with respect to a fixed prime.

We shall also consider weak uniform distribution of sequences, meaning by that uniform distribution in residue classes (mod N), prime to N. The standard example here is the sequence of all primes, which is weakly uniformly distributed (mod N) for every integer N.

After proving, in the first chapter, certain general results we shall consider uniform distribution of certain types of sequences, starting with polynomial sequences and considering also linear recurrent sequences and sequences defined by values of additive arithmetical functions. This will be done in chapter II-IV. In the last two chapter we shall study uniform distribution of sequences defined by multiplicative functions, in particular those, which are "polynomial-like", i.e. satisfy the condition $f(p^k) = P_k(p)$ for primes $p$ and $k \geq 1$ with suitable polynomials $P_1, P_2, \ldots$ . In particular we shall consider classical arithmetical functions, like the number or sum of divisors, Euler's $\varphi$-function and Ramanujan's $\tau$-function. This will lead to certain questions concerning the value distribution of polynomials.

Our tools belong to the classical number theory and include fundamentals of the theory of algebraic numbers. In certain places we shall use more recent work, like the theorems of P.Deligne, J.P.Serre and H.P.F.Swinnerton-Dyer on modular forms, which will be used in the study of Ramanujan's function. In such cases we shall explicitly state the result needed with a proper reference.

We shall use notation which is standard in number theory. In particular we shall denote the number of divisors of $n$ by $d(n)$, $\sigma(n)$ will denote the sum of divisors of n and $\sigma_k(n)$ the sum of their k-th powers, only positive divisors being taken into account. The cardinality of a set A will be denotes by $\#A$ and the letter p will be reserved for primes (except when inside a word). The ring of integers will be denoted by Z and G(N) will be the multiplicative group of restricted residue classes (mod N), i.e. the group of invertible elements of the factor ring Z/NZ. Theorems, lemmas and propositions will be consecutively numbered in each chapter. Certain open problems will be stated in the text, numbered consecutively through all chapters.

I wish to express my gratitude to Mrs Dambiec from the Department of Mathematics of Wrocław University for the patient and careful preparation of the typescript.

Wrocław, February 1984

CONTENTS

# CHAPTER I

# GENERAL RESULTS

## § 1. Uniform distribution (mod N).

1. If $N$ is a positive integer and $\{a_n\}$ a sequence of integers, then the distribution function $F(k)$ of this sequence with respect to the modulus $N$ is defined by the formula

$$F(k) = \lim_{x\to\infty} x^{-1} \# \{n \le x:\ a_n \equiv k \pmod N\} \qquad (k=0,1,\dots,N-1)$$

provided all limits here exist. In the particular case when the function $F(k)$ is constant, thus necessarily equal to $1/N$, one says that the sequence $\{a_n\}$ is *uniformly distributed* (mod N), shortly UD(mod N). Obviously this notion is a particular case of the notion of uniform distribution in compact abelian groups in the rather easy finite case, however most problems arising here have nothing to do with the abstract situation and the general approach is usually of no help.

In the sequel it will be sometimes more convenient to consider arithmetical functions $f(n)$ in place of sequences, which of course does not change the essence but presents certain advantages. We shall hence say that a function $f$ is UD(mod N) if the sequence $f(1), f(2), \dots$ of its values has this property.

From Weyl's criterion the following necessary and sufficient condition for a sequence to be uniformly distributed (mod N) results immediately, however we prefer to give a simple direct proof:

PROPOSITION 1.1. *A sequence* $\{a_n\}$ *of integers is* UD(mod N) *if and only if for* $r=1,2,\dots,N-1$ *one has*

$$\lim_{x\to\infty} x^{-1} \sum_{n\le x} \exp(2\pi i a_n r/N) = 0 \ . \qquad (1.1)$$

*Proof.* Put $f_r(t) = \exp(2\pi i t r/N)$ for $r=0,1,2,\dots,N-1$. In view of

$$\sum_{r=0}^{N-1} f_r(t) = \begin{cases} N & \text{if } N|t \\ 0 & \text{otherwise} \end{cases}$$

we get for $j=0,1,\dots,N-1$

$$x^{-1} \sum_{\substack{n\le x \\ a_n\equiv j \,(\mathrm{mod}\ N)}} 1 = (Nx)^{-1} \sum_{n\le x} \sum_{r=0}^{N-1} f_r(a_n-j) =$$

$$= (Nx)^{-1} \sum_{r=0}^{N-1} \overline{f_r(j)} \sum_{n\le x} f_r(a_n) = 1/N + (Nx)^{-1} \sum_{r=1}^{N-1} \overline{f_r(j)} \sum_{n\le x} f_r(a_n)$$

hence the condition (1.1) for $r=1,2,\dots,N-1$ is sufficient for the uniform distribution (mod N) of the sequence $\{a_n\}$. Conversely if $\{a_n\}$ is UD(mod N) then for $r=1,2,\dots,N-1$ we get

$$\sum_{n\le x} f_r(a_n) = \sum_{j=0}^{N-1} f_r(j) \sum_{\substack{n\le x \\ a_n\equiv j \,(\mathrm{mod}\ N)}} 1 = \sum_{j=0}^{N-1} f_r(j)(x/N + o(x)) =$$

$$= xN^{-1} \sum_{j=0}^{N-1} f_r(j) + o(x) = o(x)$$

since for $r=1,2,\dots,N-1$ we have obviously

$$\sum_{j=0}^{N-1} f_r(j) = 0 . \qquad \square$$

This criterion is not always easy to apply to particular sequences since it involves evaluations of exponential sums, which may be very awkward, however we shall see later that in certain cases it can be used to deduce more convenient conditions.

2. The first systematic study of uniform distribution (mod N) was done by I.NIVEN [61] and Proposition 1.1 occurs for the first time explicitly in S.UCHIYAMA [61]. Note that the assertion, contained in the last paper that uniform distribution (mod N) for all N of a sequence $\{a_n\}$ implies that $\{a_n x\}$ is UD(mod 1) for almost all x is inexact. In fact, H.G.MEIJER [70] gave a counter-example and in H.G.MEIJER, R.SATTLER [72] a sequence $\{a_n\}$ was found which is UD(mod N) for all N but $\{a_n x\}$ can be UD(mod 1) only for x from a set of measure zero. In L.KUIPERS, S.UCHIYAMA [68] the proper version

of the debated statement was proved, in which UD(mod 1) was replaced by a weaker concept.

Connections of UD(mod N) with certain measure-theoretical concepts was considered in M.UCHIYAMA, S.UCHIYAMA [62] and A.DIJKSMA, H.G.MEIJER [69]. Cf. also the book of L.KUIPERS, H.NIEDERREITER [74] (Ch. 5).

In C.L.VAN DEN EYNDEN [62] it was shown that if $\{a_n\}$ is uniformly distributed (mod 1) then for every $N \geq 1$ the sequence $[Na_n]$ is UD(mod N) and the converse also holds. The same was also proved by J.CHAUVINEAU [65], [68] who with the use of results of J.KOKSMA [35] obtained the following result, implying UD(mod N) for all N for a large class of sequences: if $f_n(x)$ is a sequence of differentiable functions on an interval I satisfying the following two conditions:

(a) If $m \neq n$, then $f_m'-f_n'$ is monotone and of constant sign on I, and

(b) There exists a positive constant C such that for $m \neq n$ one has $|f_m'-f_n'| \geq C$,

then for almost all $x \in I$ the sequence $\{[f_n(x)]\}$ is UD(mod N) for all N.

There is no difficulty in generalizing UD(mod N) to UD(mod I) where I is an ideal of an algebraic number field. This was done for the Gaussian field by L.KUIPERS, H.NIEDERREITER, J.S.SHIUE [75]. Cf. J.R.BURKE, L.KUIPERS [76], L.KUIPERS, J.S.SHIUE [80].

3. Another generalization was considered by A.F.DOWIDAR [72]: let $T = (t_{sj})$ be a triangular infinite matrix defining a regular summation method. A sequence $c_1, c_2, \ldots$ of integers is said to be T-UD(mod N) for an integer N, provided for $k=0,1,\ldots,N-1$ one has

$$\lim_{s\to\infty} \sum_{\substack{j\leq s \\ c_j \equiv k \pmod N}} t_{sj} = \frac{1}{N}$$

(see exercise 7 for a necessary and sufficient condition for T-UD(mod N)).

It follows immediately from Proposition 1.1.that a sequence $\{a_n\}$ is UD(mod N) for all integers $N \geq 2$ if and only if for every rational number r which is not an integer one has

$$\sum_{n\leq x} \exp\{2\pi i a_n r\} = o(x).$$

W.A.VEECH [71] considered a stronger condition, allowing here r to be any real number which is not an integer. Let us call the resulting property of the sequence $\{a_n\}$ Veech-UD. He called further the sequence $\{a_n\}$ well-distributed, provided for every real $r \notin Z$ one has, uniformly in $k \geq 0$,

$$\sum_{n \leq x} \exp\{2\pi i a_{n+k} r\} = o(x)$$

and proved that for sequences of the form $a_n = [P(n)]$ where $P$ is a polynomial the concepts of Veech-UD and well distribution coincide and moreover such sequence is well distributed if and only if either $P$ is linear with leading coefficient equal to the inverse of an integer or $P$ is non-linear and the additive group generated by its coefficients (except the constant term) is non-cyclic.

## § 2. The sets M(f)

One of problems arising here is to determine for a given function $f$, the set $M(f)$ of all integers $N$ such that $f$ is UD(mod $N$). We shall now prove a result which characterizes those subsets $X$ of the positive integers which have the form $X = M(f)$ for a suitable $f$.

THEOREM 1.2. (A.ZAME [72]). *If* $X$ *is a subset of the positive integers, then there exists a function* $f$ *such that* $X = M(f)$ *if and only if* $X$ *has the following property:*

*If* $n \in X$ *and* $d$ *divides* $n$, *then* $d \in X$.

*Proof*. The "only if" part is evident, since every residue class (mod $d$) is a union of $N/d$ residue classes (mod $N$). Let us thus assume that $X$ enjoys the property stated in the theorem. Write $X = \{M_1 = 1, M_2, \ldots\}$, and let $N_1 < N_2 < \ldots$ be the sequence of the remaining positive integers. For $j=1,2,\ldots$ denote by $\chi_{N_j}(x)$ the function $\exp(2\pi i x/N_j)$, which is an additive character on the integers, trivial on $N_j Z$, thus an additive character (mod $N_j$). For $n=1,2,\ldots$ let $K_n$ be the least common multiple of $M_1,\ldots,M_n,N_1,\ldots,N_n$, put

$$f_n(x) = 1 + \frac{1}{2}\sum_{j=1}^{n} (\chi_{N_j}(x) + \overline{\chi}_{N_j}(x))2^{-j}$$

and observe that $f_n(x)$ is periodic (mod $K_n$), non-negative and moreover

$$\frac{1}{K_n} \sum_{x=1}^{K_n} f_n(x) = 1$$

Assigning to an element $u$ of $Z/K_nZ$ the measure $m_n(u) = f_n(u)K_n^{-1}$ we obtain thus a non-negative measure on the group $Z/K_nZ$ which attains at the full group the value 1, hence is a probability measure. Thus it is possible to find a sequence $x_1^{(n)}, x_2^{(n)}, \ldots$ in $Z/K_nZ$ which is uniformly distributed with respect to $m_n$. This means that for every function $F(x)$ defined on $Z/K_nZ$ one has

$$\lim_{N\to\infty} \frac{1}{N} \sum_{k=1}^{N} F(x_k^{(n)}) = K_n^{-1} \sum_{u=1}^{K_n} F(u) f_n(u) . \tag{1.2}$$

It follows in particular that for $i=1,2,\ldots$ one has

$$\lim_{N\to\infty} \frac{1}{N} \sum_{k=1}^{N} \chi_{N_i}(x_k^{(n)}) = K_n^{-1} \sum_{u=1}^{K_n} \chi_{N_i}(u) f_n(u) =$$

$$= K_n^{-1} \sum_{u=1}^{K_n} \chi_{N_i}(u) + \frac{1}{2} K_n^{-1} \sum_{i=1}^{n} 2^{-i} \sum_{u=1}^{K_n} (\chi_{N_j}(u)\chi_{N_i}(u) + \overline{\chi_{N_j}(u)}\chi_{N_i}(u))$$

Since

$$\sum_{u=1}^{K_n} \chi_{N_j}(u)\chi_{N_i}(u) = \begin{cases} K_n & \text{if } N_i = N_j = 2 \\ 0 & \text{otherwise} \end{cases} ,$$

and

$$\sum_{u=1}^{K_n} \chi_{N_j}(u)\chi_{N_i}(u) = \begin{cases} K_n & N_i = N_j \\ 0 & \text{otherwise} \end{cases}$$

we get

$$\lim_{N\to\infty} \frac{1}{N} \sum_{k=1}^{N} \chi_{N_i}(x_k^{(n)}) = \frac{\varepsilon_i}{2^{i+1}} \tag{1.3}$$

where $\varepsilon_i = \begin{cases} 2 & \text{if } N_i = 2 \\ 1 & \text{otherwise} \end{cases}$

If moreover $X(x)$ is an additive character of the integers trivial on $M_iZ$ for a certain $i \le n$ then proceeding in the same way we get

$$\lim_{N\to\infty} \frac{1}{N} \sum_{k=1}^{N} X(x_k^{(n)}) = 0 . \tag{1.4}$$

Write now consecutively all non-trivial additive characters of $Z$ trivial on $M_2Z$, then those trivial on $M_3Z$, on $M_4Z$ and so forth, forming a sequence $X_1, X_2, X_3, \ldots$ of non-trivial characters. By (1.3) and (1.4) we can choose $N_0(n)$ in such way that $N_0(1) < N_0(2) < \ldots$ and for $N$ exceeding $N_0(n)$ one has for $j=1,2,\ldots,n$ the inequalities

$$\left|\frac{1}{N}\sum_{k=1}^{N} X_j(x_k^{(n)})\right| < 2^{-n} \tag{1.5}$$

and

$$\left|\frac{1}{N}\sum_{k=1}^{N} X_{N_j}(x_k^{(n)}) - \varepsilon_j 2^{-1-j}\right| < 2^{-n}, \tag{1.6}$$

and define $T(n) = 2^n N_0(1+n)$.

We claim that the sequence $x_1^{(1)}, \ldots, x_{T(1)}^{(1)}, x_1^{(2)}, \ldots, x_{T(2)}^{(2)}, \ldots, x_1^{(n)}, \ldots$ $\ldots, x_{T(n)}^{(n)}, \ldots$ has the desired properties. Let $y_n$ be its $n$-th term. If now $X$ is any additive character (mod $M_j$) for a certain $j$, then

$$\sum_{m=1}^{N} X(y_m) = \sum_{n=1}^{s}\sum_{k=1}^{T(n)} X(x_k^{(n)}) + \sum_{k=1}^{N-R} X(x_k^{(s+1)})$$

where $s = s(N)$ is defined by

$$\sum_{j=1}^{s} T(j) \le N < \sum_{j=1}^{s+1} T(j)$$

and $R = N - \sum_{j=1}^{s} T(j)$.

Using (1.5) we obtain now for $n$ large enough, say $n \ge N_1(j)$, in view of $T(n) > N_0(n)$

$$\left|\sum_{k=1}^{T(n)} X(x_k^{(n)})\right| < T(n)2^{-n}$$

and

$$\left|\sum_{k=1}^{N-R} X(x_k^{(1+s)})\right| < \begin{cases} N_0(s+1) & \text{if } N-R < N_0(s+1) \\ (N-R)2^{-s-1} & \text{otherwise} \end{cases} .$$

Observe that $(N-R)2^{-s-1}/N$ tends to zero for $N$ tending to infinity and the same applies to $N_0(s+1)/N$ because of

$$\frac{N_0(1+s)}{N} = \frac{T(s)}{2^s N} < 2^{-s} .$$

It follows that

$$\left|\frac{1}{N}\sum_{m=1}^{N} X(y_m)\right| < \frac{1}{N}\sum_{n=1}^{s} T(n)2^{-n} + o(1)$$

and since $N \geq T(1)+T(2)+\ldots+T(s)$ we get

$$\lim_{N\to\infty}\frac{1}{N}\sum_{m=1}^{N} X(y_m) = 0 .$$

The same approach shows, with the use of (1.6) that for $j=1,2,\ldots$ the ratio

$$\frac{1}{N}\sum_{m=1}^{N} \chi_{N_j}(y_m)$$

does not tend to zero, and now it suffices to apply Proposition 1.1, having in mind that every non-trivial additive character of the integers, trivial on $NZ$ is of the form $\exp(2\pi ir/N)$ with a certain integer $r$, not divisible by $N$. □

2. It was shown later by H.NIEDERREITER ([75], th.4) that if for every positive integer $N$ one prescribes a positive, normed measure $m_N$ (i.e. a sequence $a_0(N), a_1(N), \ldots, a_{N-1}(N)$ of nonnegative reals of sum 1), and assumes that if $M$ divides $N$ then for $j=0,1,\ldots,M-1$ one has

$$a_j(M) = \sum_{\substack{i \bmod N \\ i\equiv j \pmod M}} a_i(N)$$

then one can find a sequence $u_1, u_2, \ldots$ with the property that for all $N \geq 1$ and $j=0,1,\ldots,N-1$ one has

$$\lim_{x\to\infty}\frac{1}{x}\#\{n \leq x:\ u_n \equiv j \pmod N\} = a_j(N) .$$

If, in particular, $a_j(N) = 1/N$ for all $j$ and $N$, then one obtains Theorem 1.2. It is worth noting, that the sequence $\{u_n\}$ constructed by Niederreiter is always a permutation of the set of all integers.

## § 3. Weak uniform distribution (mod N)

1. We shall also study the following weaker notion of uniform distribution (mod N) in which only those terms of a given sequence are considered which are relatively prime to N. To be precise, we shall call a sequence $\{a_n\}$ *weakly uniformly distributed* (mod N) (shortly WUD mod N) provided the following two conditions are satisfied:

(i) The set $\{n: (a_n, N) = 1\}$ is infinite,

and

(ii) For every $j$ prime to $N$ one has

$$\lim_{x\to\infty} \frac{\#\{n \le x:\ a_n \equiv j \pmod N\}}{\#\{n \le x:\ (a_n, N) = 1\}} = \frac{1}{\varphi(N)} .$$

The first condition is introduced to exclude the possibility of calling a sequence like 1,2,0,0,0,....,0,0,... weakly uniformly distributed (mod 3).

The notion of weak uniform distribution (mod N) coincides with that of uniform distribution in the group G(N) of restricted residue classes (mod N) and the general theory of uniform distribution in groups leads to the following analogue of Proposition 1.1:

PROPOSITION 1.3. *If* $N$ *is a positive integer, then the sequence* $\{a_n\}$ *of integers is* WUD(mod N) *if and only if for every non-principal Dirichlet character* $\chi$(mod N), *one has*

$$\sum_{n\le x} \chi(a_n) = o\Big(\sum_{n\le x} \chi_0(a_n)\Big)$$

*where* $\chi_0$ *is the principal character* (mod N).

The proof can be carried out along the lines of that of Proposition 1.1 and we leave it to the reader. □

2. For another proof cf. S.UCHIYAMA [68], where also a relation between uniform distribution and weak uniform distribution is established. It is shown namely, that a sequence $\{u_n\}$ is UD(mod N) if and only if for every integral $c$ the sequence $\{u_n + c\}$ is weakly uniformly distributed (mod N). (The "only if" part is of course obvious).

Let us denote by $M^*(f)$ the set of all those integers $N$ for which $f$ is WUD(mod N). Again the question may be posed, how one can characterize the possible sets $M^*(f)$ in the style of Theorem 1.2. Contrary to the case of uniform distribution, here the answer is unknown and by analogy we propose the following

PROBLEM I. *Prove that if* $X$ *is a subset of the positive integers, then there exists a function* $f$ *such that* $X = M^*(f)$ *if and only if* $X$ *has the following property:*

*If* $n \in X$ *and* $d$ *is a divisor of* $n$ *which is divisible by all prime divisors of* $n$*, then* $d \in X$.

One sees immediately, that the stated condition is necessary, however the proof of its necessity seems to present difficulties. I was informed by dr I.RÚZSA, that he can prove the following result: if one divides the set of all square-free numbers into two disjoint sets A,B then it is possible to find a function $f$ which is WUD for all members of A but for no member of B. Recently Mrs. E.ROSOCHOWICZ succeeded in proving the conjecture in the case, when $X$ is an arbitrary subset of odd integers.

To the question of characterizing $M(f)$ and $M^*(f)$ for certain classes of functions we shall return in the sequel.

## § 4. Uniform distribution of systems of sequences.

1. The notion of uniform distribution (mod N) and that of weak uniform distribution (mod N) can be easily generalized to cover systems of sequences. Let $S = \langle a_n^{(1)}, \dots, a_n^{(r)} \rangle$ be such a system of $r$ sequences of integers and let $N_1, \dots, N_r$ be given positive integers. We shall say that the system $S$ is uniformly distributed with respect to $N_1, \dots, N_r$ provided for all choices of integers $c_1, \dots, c_r$ one has

$$\lim_{x\to\infty} x^{-1} \#\{n \le x:\ a_n^{(j)} \equiv c_j \pmod{N_j} \text{ for } j=1,2,\dots,r\} = 1/N_1 \dots N_r .$$

Similarly, the system S will be called weakly uniformly distributed with respect to $N_1,\dots,N_r$ if the set $\{n: (a_n^{(1)}\dots a_n^{(r)},N)=1\}$ is infinite and for every choice of integers $c_1,\dots,c_r$ satisfying $(c_j,N_j)=1$ for $j=1,2,\dots,r$ one has

$$\lim_{x\to\infty}\frac{\#\{n\le x:\ a_n^{(j)}\equiv c_j \pmod{N_j}\ \text{ for } j=1,2,\dots,r\}}{\#\{n\le x:\ (a_n^{(1)}\dots a_n^{(r)},N)=1\}}=\frac{1}{\varphi(N_1)\dots\varphi(N_r)}$$

It is clear that if S is UD with respect to $N_1,\dots,N_r$ then the sequence $a_n^{(i)}$ is UD(mod $N_i$) for $i=1,2,\dots,r$.

2. The first of these notions is a particular case of the notion of *strong asymptotic independence* considered by M.B.NATHANSON [77] and is related to the notion of *independence* (mod N) of seqnences, introduced by L.KUIPERS, J.S.SHIUE [72c] and studied later by L.KUIPERS, H.NIEDERREITER [74a] and M.B.NATHANSON [77]. We define the last notion in the case of two sequences, the passage to the general case being evident: two sequences $a_n$, $b_n$ of integers are said to be independent (mod N) if for $i,j=0,1,\dots,N-1$ all limits

$$\lim_{x\to\infty} x^{-1}\#\{n\le x:\ a_n\equiv i \pmod N\}=\alpha_i\ ,$$

$$\lim_{x\to\infty} x^{-1}\#\{n\le x:\ b_n\equiv i \pmod N\}=\beta_i\ ,$$

$$\lim_{x\to\infty} x^{-1}\#\{n\le x:\ a_n\equiv i \pmod N,\ b_n\equiv j \pmod N\}=\gamma_{ij}$$

exist and moreover one has $\gamma_{ij}=\alpha_i\beta_j$. One sees easily that a system S is UD with respect to $N,N,\dots,N$ if and only if all its members are UD(mod N) and furthermore the system is independent (mod N).

The following criterion can be proved in the same way as our Proposition 1.1:

PROPOSITION 1.4. (i) *The system* S *is* UD *with respect to* $N_1,N_2,\dots,N_r$ *if and only if for every choice of integers* $b_1,\dots,b_r$ *satisfying* $0\le b_i<N_i$ $(i=1,2,\dots,r)$ *and not all vanishing one has*

$$\sum_{n\le x}\exp\{2\pi i\sum_{j=1}^{r} a_n^{(j)}b_jN_j^{-1}\}=o(x)\ .$$

(ii) *The system* S *is* WUD *with respect to* $N_1,\dots,N$ *if and only if for every choice of characters* $\chi_i \pmod{N_i}$, $(i=1,2,\dots,r)$ *not all of them being principal, one has*

$$\sum_{n\le x} \chi_1(a_n^{(1)})\chi_2(a_n^{(2)}) \dots \chi_r(a_n^{(r)}) = o(x)$$

*and further this sum with all characters* $\chi_i$ *principal is unbounded.*

## Exercises

1. (A.DIJKSMA, H.G.MEIJER [69]). Prove that an increasing sequence $a_n$ of integers which has density 1 (i.e. $x^{-1}\#\{a_n \le x\}$ tends to unity for $x \to \infty$) is UD(mod N) for all N.

2. (L.KUIPERS, S.UCHIYAMA [68]). Prove that the sequence $[\log n]$ is not UD(mod N) for $N \ne 1$.

3. Show, that if $P(x)$ is a polynomial such that the sequence of its consecutive values is UD(mod N) for all N, then $P(x)$ must be linear.

4. Prove, that if $0 < c < 1$, then the sequence $[n^c]$ is UD(mod N) for all N. (The same holds for all $c > 0$, $c \notin Z$ ; see J.M.DESHOUILLERS [73]).

5. (C.L.VAN DEN EYNDEN [62]). Show, that if the sequence $a_n$ of real numbers is uniformly distributed (mod 1), then the sequence $[Na_n]$ is UD(mod N).

6. Prove that if $P(x)$ is a non-constant polynomial with real coefficients, at least one of which is irrational, and $P(0)-0$, then the sequence $[P(n)]$ is UD(mod N) for all N.

7. (A.F.DOWIDAR [72]). Prove, that a sequence $\{c_j\}$ of integers is T-UD(mod N) if and only if for $k=1,2,\dots,N-1$ one has

$$\lim_{s\to\infty} \sum_{j=1}^{s} t_{sj} \exp\{2\pi i c_j k/N\} = 0 .$$

8. Give an example of a polynomial P such that the sequence $[P(n)]$ is UD(mod N) for all N, without being well-distributed.

9. (I.NIVEN [61]). Prove that the sequence $[n\alpha]$ is UD(mod N) for all N if and only if either $\alpha$ is irrational, or $\alpha$ is an inverse of a non-zero integer.

# CHAPTER II

# POLYNOMIAL SEQUENCES

## § 1. Permutation polynomials

1. The simplest class of sequences which comes to mind is formed by sequences $\{P(n)\}$ of consecutive values of a polynomial $P$ with rational, integral coefficients. Since such a sequence is periodic (mod $N$) for every integer $N$, hence it will be uniformly distributed (mod $N$) if and only if the finite sequence $P(1) \bmod N, P(2) \bmod N, \ldots, P(N) \bmod N$ is a permutation of the set $1,2,\ldots,N$. If this happens, the polynomial $P$ is called a *permutation polynomial* (mod $N$). An extensive study of such polynomials in the case of prime $N$ was initited by L.E.DICKSON in his thesis ([97]) where he amplified and vastly extended previous observations of E.BETTI [51],[52],[55], E.MATHIEU [61], C.HERMITE [63] and others.

2. Dickson considered, more generally, permutation polynomials in arbitrary finite fields, however he was never seriously interested in permutation polynomials (mod $N$) for $N$ composed. He had reasons for that. In fact, the next simple result shows that in studying permutation polynomials one can restrict attention to prime moduli:

PROPOSITION 2.1. (W.NÖBAUER [65]). (i) *A polynomial* $P(x)$ *over* $Z$ *is a permutation polynomial* (mod $N$), *for* $N = \prod_p p^{a_p}$ *if and only if it is a permutation polynomial* (mod $p^{a_p}$) *for every prime* $p$ *dividing* $N$.

(ii) *A polynomial* $P(x) \in Z[x]$ *is a permutation polynomial* (mod $p^2$) *with a prime* $p$ *if and only if it is a permutation polynomial* (mod $p$) *and the congruence* $P'(x) \equiv 0 \pmod p$ *has no solutions.*

(iii) *If* $P(x) \in Z[x]$ *is a permutation polynomial* (mod $p^2$) *with a prime* $p$, *then it is also a permutation polynomial for all powers of* $p$.

*Proof.* (i) The "only if" part is a consequence of the easy part of Theorem 1.2. To prove the "if" part it is sufficient to show that if $P(x)$ is a permutation polynomial (mod $M$) and (mod $N$) with coprime $M$ and $N$ then it is also such with respect to $MN$. However if $(M,N)=1$, then the Chinese Remainder Theorem implies that if the maps $Z/MZ \to Z/MZ$ and $Z/NZ \to Z/NZ$ induced by $P$ are both surjective then the induced map $Z/(MN)Z \to Z/(MN)Z$ is also surjective and hence must be a permutation.

(ii) If $P'(x) \equiv 0 \pmod p$ is insolvable, and $P(x)$ is a permutation polynomial (mod $p$), then for every $m$ the congruence $P(x) \equiv$ $\equiv m \pmod{p^2}$ has a unique solution, thus $P(x)$ is a permutation polynomial (mod $p^2$). If however $P'(x_0) \equiv 0 \pmod p$ then the congruence $P(x) \equiv P(x_0) \pmod{p^2}$ has $p$ solutions, thus $P$ is not a permutation polynomial (mod $p^2$).

(iii) From the assumptions it follows by (ii) that $P'(x) \equiv 0 \pmod p$ has no solutions and thus for every $m$ the unique solution of $P(x) \equiv$ $\equiv m \pmod{p^2}$ can be lifted to a unique solution of $P(x) \equiv m \pmod{p^k}$ for $k=3,4,\ldots$ . □

COROLLARY. *For a given polynomial* $P \in Z[X]$ *denote by* $S(P)$ *the set of all primes* $p$ *such that* $P$ *is a permutation polynomial* (mod $p^2$) *and by* $T(P)$ *the set of all primes* $p$ *such that* $P$ *is a permutation polynomial* (mod $p$) *but not* (mod $p^2$). *Then the set* $M(P)$ *coincides with the set of all integers of the form* $p_1 \ldots p_r q_1^{a_1} \ldots q^{a_s}$ *with* $r \geq 0$, $s \geq 0$; $p_1,\ldots,p_r \in T(P)$; $q_1,\ldots,q_s \in S(P)$; $a_j > 0$.

*Proof.* Follows immediately from the proposition. □

The Theorem 2.8 below will show that for $S(P)$ and $T(P)$ one can take arbitrary finite sets of primes. The following question arises thus:

PROBLEM II. *For which disjoint sets* $S,T$ *of primes one can find a polynomial* $P$ *such that* $S(P)=S$ *and* $T(P)=T$ *?*

It seems that one should be able to deduce an answer to it using Fried's solution of a problem of Schur which we shall quote in section 4 below.

## § 2. Generators for the group of permutation polynomials

1. If $p$ is a prime number, then in view of the congruence $x^p \equiv x \pmod p$ holding identically in $x$, one can restrict the attention to polynomials of degree not exceeding $p-1$. Having this in mind it is not difficult to obtain a description of all permutation polynomials $\pmod p$:

THEOREM 2.2. (L.CARLITZ [53]). *If* $p$ *is a prime and* $P(x)$ *is a permutation polynomial* $\pmod p$, *then there exists a polynomial* $Q(x)$ *such that for all* $x$ *we have* $P(x) \equiv Q(x) \pmod p$ *and the polynomial* $Q(x)$ *is a composition of polynomials of the form* $ax+b$ *(with* $a \not\equiv 0 \pmod p$) *and the polynomial* $x^{p-2}$

*Proof.* It suffices to show that every transposition $(0t)$ (for $t=1,2,\dots,p-1$) of the set $\{0,1,\dots,p-1\}$ is induced by a polynomial $Q(x)=Q_t(x)$ satisfying the assertion of the theorem. Writing $u'$ for the inverse of the element $u \pmod p \neq 0$ we can define such a polynomial by the formula

$$Q_t(x) = -t^2((x-t)^{p-2}+t')^{p-2}-t)^{p-2} .$$

Clearly $Q_t$ is a composition of linear polynomials and the polynomial $x^{p-2}$ and it is also obvious that $Q \pmod p$ interchanges the element $0$ and $t$. If $x \neq 0,t$ then

$$Q_t(x) = -t^2((x-t)'+t')^{p-2}-t)^{p-2} \equiv -t^2(-t^2x')^{q-2} \equiv x \pmod p$$

proving that $Q_t$ satisfies our needs. □

(Note that this argument holds without change for the analogue of the assertion in arbitrary finite fields. L.CARLITZ [53]).

Earlier the case $p=5$ was considered by E.BETTI [51] and the case $p=7$ by L.E.Dickson [97].

2. There are other possible variants of this theorem. So K.D.FRYER [55] showed that one can replace the set $\{ax+b:\ a \not\equiv 0 \pmod p\} \cup \{x^{p-2}\}$ by the pair $\{x+1,\ ax^{p-2}\}$ where a is any fixed integer which is a quadratic residue $\pmod p$ in case $p \equiv 1 \pmod 4$ and a quadratic non-

residue in case $p \equiv 3 \pmod 4$. C.WELLS [68] proved that the same role may be taken by the triple $\{x+1, x^{p-2}, gx\}$ where $g$ is a primitive root $\pmod p$.

Choosing adequately sets of permutation polynomials $\pmod p$ one can often give a simple set of generators for various subgroups of the full symmetric group on $p$ letters. Examples of this procedure can be found already in the classical work of L.E.DICKSON [97],[01]. From newer papers on this subject the reader may consult H.HULE, W.B.MÜLLER [73], H.LAUSCH, W.B.MÜLLER, W.NÖBAUER [73], R.LIDL [73], R.LIDL, W.B. MÜLLER [76], C.WELLS [67].

## § 3. Hermite's characterization of permutation polynomials

1. The following result of C.HERMITE [63] is often helpful in proving that a particular polynomial is not permutational.

THEOREM 2.3. *Let* $p$ *be a prime and* $P(x)$ *a polynomial over* $Z$ *with its degree not exceeding* $p-1$. *Then* $P(x)$ *will be a permutation polynomial* $\pmod p$ *if and only if for every* $t=1,2,\ldots,p-1$ *there exists a polynomial* $Q_t(x)$ *of degree* $\le p-2$ *such that the congruence*

$$P^t(a) \equiv Q_t(a) \pmod p$$

*holds for every a and moreover the congruence* $P(x) \equiv 0 \pmod p$ *has exactly one solution.*

(A variant of this theorem was proved by L.CARLITZ, J.A.LUTZ [78]).

*Proof*. Observe first that one has

$$\sum_{x=0}^{p-1} x^j \equiv \begin{cases} 0 \pmod p & \text{for } j=1,2,\ldots,p-2 \\ -1 \pmod p & \text{for } j=p-1 \end{cases} \tag{2.1}$$

Indeed, if $g$ is a primitive root $\pmod p$, then the sum on the left hand-side equals

$$\sum_{k=1}^{p-1} g^{kj}$$

and the assertion follows immediately.

For any polynomial $P(x) = \sum_{i=0}^{p-1} a_i x^i$ and $t=1,2,\dots,p-2$ write $P^t(x) = \sum_{i=0}^{p-1} a_{i,t} x^i$. Adding these equalities and utilizing (2.1) we obtain for every $b$

$$\sum_{b=0}^{p-1} P^t(b) \equiv \sum_{i=0}^{p-1} a_{i,t} \sum_{b=0}^{p-1} b^i \equiv -a_{p-1,t} \pmod p \tag{2.2}$$

If now $P$ is a permutation polynomial $\pmod p$, then in view of $1 \le t \le p-2$

$$\sum_{b=0}^{p-1} P^t(b) = \sum_{b=0}^{p-1} b^t \equiv 0 \pmod p$$

thus $a_{p-1,t} \equiv 0 \pmod p$.

Noting that a permutation polynomial has exactly one zero $\pmod p$ we obtain the necessity of our conditions. To prove their sufficiency note that due to (2.2) they imply the congruences

$$\sum_{b=0}^{p-1} P^t(b) \equiv 0 \pmod p \tag{2.3}$$

for $t=1,2,\dots,p-2$ and

$$\sum_{b=0}^{p-1} P^{p-1}(b) \equiv -1 \pmod p .$$

Let $V(x) = \sum_{j=0}^{p} A_j x^j$ $(A_p=1)$ be the polynomial over the finite field of $p$ elements, whose roots are the residues mod $p$ of $P(0), P(1), \dots, P(p-1)$. Using Newton's formulas we obtain, with $S_t = \sum_{b=0}^{p-1} P^t(b) \pmod p$, the equalities

$$S_k + \sum_{i=1}^{k-1} A_{p-i} S_{k-i} + kA_{p-k} = 0 \qquad (k=1,2,\dots,p)$$

thus in view of (2.3) for $k=1,2,\dots,p-2$ the equality $kA_{p-k}=0$ results, showing that $A_2=\dots=A_{p-1}=0$ and $V(x)=x^p+A_1x+A_0$. Since $P(0)\equiv 0 \pmod p$ we get $A_0=0$. Moreover for $j=0,1,\dots,p-1$ we have

$0 = V(P(j)) = P^p(j) + A_1P(j) = (1+A_1)P(j)$.

If we would have $A_1 \not\equiv -1$ then the congruence

$$0 \equiv P(0) \equiv P(1) \equiv \dots \equiv P(p-1) \pmod p$$

would result, contradicting our assumption. Thus $A_1=-1$ and we see finally, that the sequence $P(0) \bmod p, \dots, P(p-1) \bmod p$ coincides with the set of all roots of $x^p-x$ in $GF(p)$ and hence forms a permutation of $GF(p)$. Consequently $P$ is a permutation polynomial $\pmod p$. □

As shown by L.J.ROGERS [91] one can replace the condition $1\le t\le p-1$ in the theorem by $1\le t\le (p-1)/2$. The same observation was later made by V.A.KURBATOV, N.G.STARKOV [65].

2. COROLLARY. *If* $k\ge 2$ *and* $p$ *is a prime congruent to* $1 \pmod k$ *then a polynomial of degree* $k$ *with its leading coefficient not divisible by* $p$ *cannot be a permutation polynomial* $\pmod p$

*Proof.* It suffices to consider monic polynomials, so let $V(x) = x^k+a_{k-1}x^{k-1}+\dots+a_0$ be a permutation polynomial $\pmod p$ and assume that $p\equiv 1 \pmod k$. Then $V(x)-a_0$ is again a permutation polynomial $\pmod p$. Putting $m=(p-1)/k$ we get

$$(V(x)-a_0)^m = x^m(x^{k-1}+a_{k-1}x^{k-2}+\dots+a_1)^m = x^m(x^{m(k-1)}+\dots+a_1^m)$$

and since $m(k-1)<p$ and the coefficient of $x^{p-1}=x^m$ of $(V(x)-a_0)^m$ equals unity, we get a contradiction with the theorem. □

This corollary implies that for a polynomial $P$ of degree 2 the set $M(P)$ coincides with the set of integers whose prime factors divide the leading term of $P$ but do not divide the middle term. In particular $M(P)$ contains in this case only a finite number of primes. DICKSON's results [97] imply that for quartic and sextic polynomials the same assertion is true. S.R.CAVIOR [63] posed the question, whether the same fact holds for every polynomial of an even degree. We shall see later, that this is in fact true. (See theorem 2.7).

## § 4. Examples

1. Now we shall produce examples of permutation polynomials. Clearly the linear polynomial $ax+b$ is permutational for every prime $p$ not dividing $a$. It is also obvious that a quadratic polynomial $ax^2+bx+c$ is permutational (mod $p$) for an odd prime $p$ if and only if $p|a$, $p\nmid b$. Indeed, if $p$ does not divide $a$, then our polynomial is congruent (mod $p$) to a polynomial of the form $(Ax+B)^2+C$ and so its image contains at most $1+(p-1)/2$ residues (mod $p$). One sees also immediately, that the binomial polynomial $ax^n+b$ is a permutation polynomial (mod $p$) for every prime $p\nmid a$ satisfying $(n,p-1)=1$. L.E.DICKSON [97] characterized polynomials of degrees $\le 6$ which are permutational with respect to at least one prime, and determined also the appropriate primes. The reader may rediscover his result as an exercise.

2. To give non-trivial examples we shall now consider the Čebyšev polynomials. They can be defined in various manners, the simplest being the following: for $n=1,2,\ldots$ and $-1\le x\le 1$ put

$$T_n(x) = \cos(n \arccos x) .$$

LEMMA 2.4. *For* $n\ge 1$ *the function* $T_n(x)$ *can be extended to a polynomial with integral coefficients, of degree* $n$ *and leading coefficient* $2^{n-1}$.

*Proof.* Let $f_n(x) = \sin(n \arccos x) = (1-x^2)^{\frac12} U_n(x)$. We claim that for all $n\ge 1$ the following facts hold:

(i) $T_n(x)$, $U_n(x)$ can be extended to polynomials over $Z$,

(ii) $\deg T_n = n$, $\deg U_n = n-1$

and

(iii) The leading coefficients of $T_n$ and $U_n$ are equal to $2^{n-1}$.

Since $T_1(x)=x$ and $U_1(x)=1$ (i) - (iii) are true for $n=1$. Using the identity

$$f_{n+1}(x) = xf_n(x) + (1-x^2)^{\frac12} T_n(x) ,$$

we obtain

$$U_{n+1}(x) = T_n(x) + xU_n(x) \tag{2.4}$$

and

$$T_{n+1}(x) = xT_n(x) - f_1(x)f_n(x) = xT_n(x) - x^2U_n(x) - U_n(x)$$

and this leads to (i) - (iii) by an easy recurrence argument. □

We shall have no opportunity to use neither the explicit form

$$T_n(x) = 2^{n-1}x^n + \sum_{k=1}^{[n/2]} (-1)^k\{\binom{n-k}{k} + \binom{n-k-1}{k-1}\}2^{n-2k}x^{n-2k}$$

for Čebyšev polynomials (which can be proved by recurrence) nor many other fascinating properties of them, referring the interested reader either to any book treating orthogonal polynomials or to I.SCHUR [73] which paper is a source of many arithmetical results concerning $T_n(x)$ and $U_n(x)$.

The only further property of $T_n$ we need is given in the next lemma. It may be used to give a definition of $T_n$ independent on trigonometric functions.

LEMMA 2.5. *Let* $x$ *be a real number and let* $u = u(x) = x + (x^2-1)^{\frac{1}{2}}$, $v = v(x) = x - (x^2-1)^{\frac{1}{2}}$ *be the roots of* $Y^2 - 2xY + 1$. *Then*

$$T_n(x) = \frac{u^n + v^n}{2}$$

*and*

$$U_n(x) = \frac{u^n - v^n}{2(x^2-1)^{\frac{1}{2}}}$$

*Proof*. For $n=1$ the assertion results immediately from $T_1(x) = x$, $U_1(x) = 1$ and the equalities (2.4) and (2.5) pave the way for an easy recurrence argument. □

3. Now we can prove the main result of this section obtained first by DICKSON [97]:

THEOREM 2.6. *If* $p$ *is an odd prime number, then the polynomial* $T_n$ *is a permutation polynomial* (mod p) *if and only if* $(p^2-1,n)=1$.

*Proof.* To prove the sufficiency we use an elegant argument due to M.FRIED [70]. (Dickson's original proof concentrated rather on surjectivity than on injectivity of $T_n$). We start with the observation, resulting from the last lemma, that if $x$ is an integer and $\bar{x}$ the corresponding residue (mod p) treated as an element of the field $GF(p)$, then the residue $T_n(x) \bmod p = A_n(x)$ equals

$$\frac{1}{2}(t^n(x) + t^{-n}(x))$$

where $t(x)$ is for every $x$ an element of $GF(p^2)$ satisfying $t^2 - 2\bar{x}t + 1 = 0$.

In particular

$$t(x) + t^{-1}(x) = 2\bar{x} . \tag{2.6}$$

Let $(n,p^2-1)=1$, $x_1 \not\equiv x_2 \pmod p$, and denote for $i=1,2$ by $t_i$ one of the solutions of

$$2\bar{x}_i = y + y^{-1}$$

in $GF(p^2)$. If $T_n(x_1) \equiv T_n(x_2) \pmod p$, then $A_n(x_1) = A_n(x_2)$, hence

$$t_1^n + t_1^{-n} = t_2^n + t_2^{-n} = c ,$$

with a certain $c \in GF(p)$. This shows that $t_1^n, t_1^{-n}$, $t_2^n, t_2^{-n}$ all satisfy the equation $y^2 - cy + 1$ and thus we must have either $t_1^n = t_2^n$ or $t_1^n = t_2^{-n}$. Since the multiplicative group of $GF(p^2)$ is cyclic of $p^2-1$ elements, it follows that either $t_1 = t_2$ or $t_1 = t_2^{-1}$ must hold, which by (2.6) implies $\bar{x}_1 = \bar{x}_2$, a contradiction.

To prove the necessity observe first that in the case $(n,p-1) \neq 1$ the polynomial $T_n$ cannot be permutational (mod p) because of the Corollary to Theorem 2.3. Thus assume $(n,p+1) \neq 1$.

From the deliberations above we keep in mind that in order to show that $T_n$ is not a permutation polynomial (mod p) it suffices to find integers $x_1,x_2$ incongruent (mod p) for which one would have $t^n(x_1) = t^n(x_2)$. Since

$T_{rs}(x) = T_r(T_s(x))$ we may assume that n is an odd prime divisor of 1+p. Expressing $\bar{x}$ in terms of t(x) we obtain immediately that the image of t(x) consists of those elements of $GF(p^2)$ for which $(u^{p+1}-1)(u^{p-1}-1)=0$. There are p+1 elements $u \in GF(p^2)$ with $u^{p+1} = 1$ all lying in the image of t(x) and we obtain that there are at most (p+1)/n possibilities for $u^n$, thus there must be a pair $x_1 \not\equiv x_2 \pmod p$ such that $t^n(x_1) = t^n(x_2)$. □

COROLLARY. *If* $P(x) \in Z[x]$ *is a composition of Čebyšev polynomials* $T_{n_1},\dots,T_{n_s}$ *with* $(n_j,6)=1$ *for* $j=1,2,\dots,s$ *and of polynomials* $a_jx^{m_j}+b_j$ *(with odd* $m_j$, *non-zero integral* $a_j$*'s and arbitrary integral* $b_j$*'s for* $j=1,2,\dots,r$), *in an arbitrary order, then* P(x) *is a permutation polynomial for infinitely many primes* p.

*Proof.* It suffices to prove that there are infinitely many primes p satisfying $(n_j,p^2-1)=1$ for $j=1,2,\dots,s$ and $(m_j,p-1)=1$ for $j=1,2,\dots,r$, but this follows from Dirichlet's prime number theorem applied to the progression $N \equiv 2 \pmod{n_1\dots n_s m_1\dots m_r}$. □

## § 5. Consequences of Fried's theorem

1. The corollary to the Theorem 2.6 provides us with polynomials which are permutation polynomials for infinitely many primes. It was proved by I.SCHUR [23] that any polynomial of prime degree which is permutational for infinitely many primes must be a composition of binomials $ax^n+b$ and Čebyšev polynomials. He called an integer N a *Dickson number*, if the same assertion is true for all polynomials of degree N and stated also a theorem, whose proof he promised to publish later, giving a sufficient condition for N to be a Dickson number in terms of finite permutation groups. Although in a later paper (I.SCHUR [33]) he was able to show that this condition holds for every N and promised to give arithmetical applications of this result, he never returned to this subject. Since then, the assertion that every integer is a Dickson number was commonly called *Schur's conjecture*.

After certain new classes of Dickson numbers were found by V.A. KURBATOV [49] and U.WEGNER [28] the truth of Schur's conjecture was established by M.FRIED [70], who utilized the theory of Riemann surfaces. These methods lying outside the scope of these lectures we do not present a proof of Fried's result, and limit ourselves to a deduction of two corollaries of it. The first answers a question posed by S.R.CAVIOR which we quoted in § 3:

THEOREM 2.7. (H.DAVENPORT, D.J.LEWIS [63]). *If* $P(x)$ *is a polynomial over* $Z$ *of even degree then it can be a permutation polynomial* (mod p) *only for a finite number of primes* $p$*, i.e. the set* $M(P)$ *cannot contain infinitely many primes.*

*Proof*. If the set $M(P)$ contains infinitely many primes, then by Fried's theorem $P(x)$ is a composition of polynomials $P_1(x),\dots,P_r(x)$ each of them being equal either to a certain Čebyšev polynomial or to a binomial $ax^n+b$. Since from the definition of Čebyšev polynomials it follows that for all $m,n$ one has $T_{mn}(x)=T_m(T_n(x))$ hence we may assume that all polynomials $T_n(x)$ which occur in the set $\{P_1,\dots,P_r\}$ have prime indices. Since the degree of $P$ equals $\prod_{i=1}^{r}\deg P_i$ it follows that at least one $P_j$ must be of even degree, hence it equals either $T_2(x)=2x^2-1$ or $ax^n+b$ with a certain even $n$. In both cases it cannot be permutational for infinitely many primes and so the intersection $\bigcap_{i=1}^{r} M(P_i)$ is finite. However this intersection equals $M(P)$ and we arrive at a contradiction. □

Certain special cases of Theorem 2.7 were known earlier. For polynomials of degrees n=2, 4 or 6 it was shown by L.E.DICKSON [97], the case n=10 was settled by D.R.HAYES [67] and the case when $n$ is a power of two was treated in H.LAUSCH, W.NÖBAUER [73].

2. Another consequence of Fried's theorem shows that the sets $S(P)$ and $T(P)$ occuring in the corollary to Proposition 2.1 can be arbitrary disjoint finite sets of primes.

THEOREM 2.8. (W.NÖBAUER [66]). *Let* $S,T$ *be finite and disjoint sets of primes. Then there exist infinitely many polynomials* $P(x)$ *over* $Z$ *with the property, that an integer* $N$ *belongs to* $M(P)$ *if and only if* $N$ *has no prime divisors outside the union* $S\cup T$ *and is not divisible by a square of any prime from* $T$.

*Proof.* If $S \cup T$ is empty, then the polynomial $P(x) = 2x^2$ will do. Otherwise let $q$ be any prime which exceeds all elements of $S \cup T$ and define for $p$ in $S \cup T \cup \{q\}$

$$P_p(x) = \begin{cases} x & \text{if } p \in S \\ x^{a_p} & \text{if } p \in T \\ x^{2a} & \text{if } p = q \end{cases}$$

where $a_p$ is an arbitrary prime not dividing $p(p-1)$ and $a$ is an arbitrary integer exceeding $1+a_p$ for all $p \in T$. Let $P_0(x)$ be a polynomial of degree $2a$ which satisfies

$$P_0(x) \equiv P_p(x) \pmod p$$

for all $p \in S \cup T \cup \{q\}$. By construction $P_0$ is a permutation polynomial (mod p) for $p \in S \cup T$ and Proposition 2.1 shows that $M(P_0)$ contains also all powers of $p \in S$ but does not contain $p^2$ for $p \in T$. By Theorem 2.7 $M(P_0)$ can contain only finitely many primes, hence if we denote by $D$ the product of those of them which do not lie in $S \cup T$ and put $P(x) = DP_0(x)$ then the Corollary to Proposition 2.1 will show that $P(x)$ satisfies our needs. Varying $a$ we can obtain in this way polynomials of arbitrary large degree, satisfying our assertion. □

## § 6. Weak uniform distribution (mod N) of polynomials

As far we considered only uniform distribution (mod N), so let us look now at weak uniform distribution (mod N) of polynomial sequences. The following simple result reduces the problem to the study of permutation polynomials (mod p):

PROPOSITION 2.9. (i) *A polynomial* $P(x)$ *over* $Z$ *is* WUD(mod N) *for* $N = \prod_f p^{a_p}$ *if and only if it is* WUD(mod $p^{a_p}$) *for each prime* $p$ *dividing* $N$.

(ii) *A polynomial* $P(x) \in Z[x]$ *is* WUD(mod $p^k$) *for a prime power* $p^k$ *if and only if it is a permutation polynomial* (mod p), *and the congruence* $P'(x) \equiv 0 \pmod p$ *has no solutions* $x_0$ *satisfying* $p \nmid P(x_0)$.

*Proof.* (i) It is enough to show that if $(M,N)=1$ then $P(x)$ is WUD(mod $MN$) if and only if it is WUD(mod $M$) and WUD(mod $N$). If $P(x)$ is WUD(mod $MN$) and we denote by $A(T)$ the number of $x \pmod T$ such that $(T,P(x))=1$, then obviously $A(MN)=A(M)A(N)$ and so, if $(i,M)=$ $=(j,N)=1$, then

$$\#\{1\le x\le MN;\ P(x)\equiv i \pmod M,\ P(x)\equiv j \pmod N\} = \frac{A(M)A(N)}{\varphi(M)\varphi(N)},$$

hence

$$N\,\#\{x \bmod M:\ P(x)\equiv i \pmod M\} = \#\{x \bmod MN:\ P(x)\equiv i \pmod M\} =$$

$$= \sum_{\substack{j \bmod N\\ (j,N)=1}} \#\{x \bmod MN:\ P(x)\equiv i \pmod M,\ P(x)\equiv j \pmod N\} +$$

$$+ \sum_{\substack{j \bmod N\\ (j,N)\ne 1}} \#\{x \bmod MN:\ P(x)\equiv i \pmod M,\ P(x)\equiv j \pmod N\} =$$

$$= \frac{A(M)A(N)}{\varphi(M)} + \sum_{\substack{j \bmod N\\ (j,N)\ne 1}} \#\{x \bmod M:\ P(x)\equiv i \pmod M\}\cdot\#\{x \bmod N:\ P(x)\equiv j \pmod N\} =$$

$$= \frac{A(M)A(N)}{\varphi(M)} + \#\{x \bmod M:\ P(x)\equiv i \pmod M\}\ (N-A(N))$$

and so

$$\#\{x \bmod M:\ P(x)\equiv i \pmod M\} = \frac{A(M)}{\varphi(M)}$$

showing that $P(x)$ is WUD(mod $M$). Interchanging $M$ and $N$ we obtain the same assertion for $N$.

Now assume that $P(x)$ is WUD(mod $M$) and WUD(mod $N$). If $(i,M)=$ $=(j,N)=1$, then

$$\#\{x \bmod MN:\ P(x)\equiv i \pmod M,\ P(x)\equiv j \pmod N\} =$$

$$= \#\{x \bmod M:\ P(x)\equiv i \pmod M\}\ \#\{x \bmod N:\ P(x)\equiv j \pmod N\} =$$

$$= \frac{A(M)A(N)}{\varphi(M)\varphi(N)} = \frac{A(MN)}{\varphi(MN)}$$

thus $P(x)$ is WUD(mod MN).

(ii) If $P(x)$ is WUD(mod $p^k$), then it is also WUD(mod p), thus among $P(1),\dots,P(p)$ every value $a \pmod p \neq 0$ occurs once, so $0 \pmod p$ must also occur one time, and UD(mod p) follows. After this observation one needs only to repeat the proof of the corresponding part of Proposition 2.1 to obtain our assertion. □

## § 7. Notes and comments

1. To conclude the story of uniform distribution of polynomial sequences let us mention certain related results and generalizations.

A polynomial $f$ over $Z$ is called *exceptional* ((or *virtually-one--one*) with respect to the prime $p$, if the polynomial $(f(x)-f(y))/(x-y)$, when factored into irreducibles (mod p) has no factor which is *absolutely irreducible*, i.e. remains irreducible in the algebraic closure of GF(p). H.DAVENPORT, D.J.LEWIS [63] conjectured that if $f \in Z[x]$, $p$ is sufficiently large prime and $f$ is exceptional with respect to $p$, then $f$ is a permutation polynomial (mod p). This conjecture was proved by C.R.MAC CLUER [66]. Another proof was given by K.S.WILLIAMS [68]. (Cf. also H.LAUSCH, W.NÖBAUER [73], th. 8.31). It was later shown by S.D.COHEN [70] that the restriction on $p$ is unnecessary, thus exceptional polynomials are permutation polynomials for all primes $p$.

In certain cases Cohen's result can be inverted, in fact, if $f(x)$ is a non-linear permutation polynomial (mod p), and $p$ is sufficiently large, say $p \geq C(\deg f)$, then $f$ is exceptional with respect to $p$. This was established by D.R.HAYES [67] and a proof is also given in H.LAUSCH, W.NÖBAUER [73] (th. 8.81).

For a generalization of the notion of permutation polynomials to rings of algebraic integers see H.NIEDERREITER, S.K.LO [79].

2. One can also generalize the notion of a permutation polynomial (mod p) to the case of several variables. A sequence $f_1,\dots,f_n$ of $n$ polynomials over $Z$ in $n$ variables is called a *permutation vector* (mod p), provided the map $[x_1,\dots,x_n] \to [f_1(x_1,\dots,x_n),\dots,f_n(x_1,\dots,x_n)]$ of $Z^n$ to $Z^n$ induces a one-to-one mapping of $(Z/pZ)^n$ onto itself. Every component of a permutation vector is called a *permutation polynomial.* This generalization was first considered by W.NÖBAUER [64], who

dealt with polynomials over arbitrary rings. This topic was pursued by several authors. We mention here R.LIDL [71], who characterized polynomial vectors in two variables over a finite field of odd characteristic and in the same case described quadratic permutation polynomials. The last result, without restriction on the characteristic was also obtained by H.NIEDERREITER [72b]. In R.LIDL [72] the analogue of Theorem 2.2 was obtained for permutation vectors in two variables over an arbitrary finite field. Cf. also R.LIDL, H.NIEDERREITER [72], R.LIDL, C. WELLS [72].

3. Another generalization was considered by J.V.BRAWLEY, L.CARLITZ, J.LEVINE [75], who proved that a polynomial over a finite field $K = GF(q)$ induces a permutation of the ring of $n\times n$ matrices over $K$ if and only if is a permutation polynomial in $GF(q), GF(q^2), \dots, GF(q^n)$ and its derivative does not vanish in $GF(q^N)$ where $N = n/2$ . Later J.V.BRAWLEY [76] solved the analogous problem in the case when $GF(q)$ is replaced by an arbitrary finite commutative ring with unit.

4. Recently H.NIEDERREITER and K.H.ROBINSON [82] considered *complete mapping polynomials* over finite fields, defined as those permutation polynomials $P(x)$ for which the polynomial $P(x)+x$ is also permutational. They determined all such polynomials of degrees smaller than 6 and produced several examples.

5. Finally let us note that the study of uniform distribution of system of polynomials with respect to a set of moduli as defined in ch. I. § 4 is uninteresting, since one sees easily that a system $P_1, \dots, P_r$ of polynomials is UD with respect to $N_1, \dots, N_r$ if and only if for $i=1,2,\dots,r$ the polynomial $P_i$ is a permutation polynomial $\pmod{N_i}$ and the moduli $N_1, \dots, N_r$ are pairwise relatively prime. An analogous statement holds for weak uniform distribution.

## Exercises

1. (L.CARLITZ [62a]). Prove that if $p$ is a sufficiently large prime $\equiv 1 \pmod 3$ then the polynomial $x^{(p+2)/3} + ax$ is a permutation polynomial $\pmod p$ for at least one choice of $a$.

2. (L.CARLITZ [62a]). Prove that $p$ is a prime $\geq 7$, then with a suitable choice of $a \not\equiv 0 \pmod p$ the polynomial $x^{(p+1)/2} + ax$ is a permutation polynomial (mod p).

3. (L.CARLITZ [63]). Prove that for any prime $p$ with a suitable choice of $a$ the polynomial $x^{(p-1)/2} + ax$ is a permutation polynomial (mod p).

4. Determine all cubic and quartic polynomials which are permutation polynomials for suitable primes.

5. (L.CARLITZ [62b]). Prove that if $r(x)$ is a rational function (mod p) which induces a permutation of the set of residue classes (mod p) enlarged by an element $\infty$ with the usual conventions $(1/0 = \infty$, $1/\infty = 0$, $\infty + a = \infty$, $a\infty = \infty$ for $a \neq 0)$ then there exists a permutation polynomial $f$ (mod p) such that one has

$$r(x) = f((Ax+B)/(Cx+D))$$

with suitable $A, B, C, D$ satisfying $\det\begin{pmatrix} A & B \\ C & D \end{pmatrix} \not\equiv 0 \pmod p$.

6. (K.D.FRYER [55]). Prove that every permutation of the field of $p \neq 2$ elements is a superposition of the polynomials $x+1$ and $ax^{p-2}$ where $a$ is a quadratic residue (mod p) in case $p \equiv 1 \pmod 4$ and a non-residue otherwise.

7. (H.NIEDERREITER, K.H.ROBINSON [82]). A polynomial $P(x)$ is called a complete mapping polynomial (mod p) if both $P(x)$ and $x + P(x)$ are permutation polynomials (mod p). Prove that if $p \neq 2,3$ then there are no such polynomials of degree $p-1$ and $p-2$, but for certain primes $p$ there exist some with degree $p-3$.

8. (loc. cit.). Prove that there do not exist complete mapping polynomials of degree 2.

## CHAPTER III

## LINEAR RECURRENT SEQUENCES

### § 1. Principal properties

1. Another class of sequences for which uniform distribution was studied consists of *linear recurrent sequences* $\{u_n\}$, defined by the condition

$$u_{n+k} = a_0 u_n + a_1 u_{n+1} + \dots + a_{k-1} u_{n+k-1} \qquad (n \geq 0) \tag{3.1}$$

where $k$ is a fixed positive integer, called the *order of the sequence* $\{u_n\}$ and $a_0, a_1, \dots, a_{k-1}$ are fixed numbers.

The oldest example of such a sequence goes back to the thirteenth century and is usually associated with rabbit breeding. It is the celebrated *Fibonacci sequence* defined by $F_{n+2} = F_n + F_{n+1}$ and $F_0 = F_1 = 1$ (or sometimes by $F_o = 0$, $F_1 = 1$, in which case the indices become shifted by 1).

To every sequence satisfying (3.1) there corresponds the polynomial

$$f(x) = x^k - a_{k-1}x^{k-1} - \dots - a_0$$

called *the polynomial associated with* $\{u_n\}$. It is obvious that the sequence $\{u_n\}$ is completely determined by its associated polynomial and the initial values $u_0, u_1, \dots, u_{k-1}$.

It is not difficult to find an explicit formula for the general term of a linear recurrent sequence:

LEMMA 3.1. *Let* $\{u_n\}$ *be a sequence of rational integers satisfying* (3.1) *and assume that its associated polynomial* $f(x)$ *has rational integral coefficients. Assume further that*

$$f(x) = \prod_{j=1}^{r} (x - \alpha_j)^{e_j}$$

*is its canonical factorization, with* $\alpha_1,\ldots,\alpha_r$ *distinct and denote by* K *the splitting field of* f *over the rationals. Then the following holds:*

(i) *There exist polynomials* $P_1,\ldots,P_r$ *with coefficients in* K, *with* $\deg P_i \le e_i - 1$ $(i=1,2,\ldots,r)$ *such that for* $n=0,1,2,\ldots$ *the equality*

$$u_n = \sum_{i=1}^{r} P_i(n)\alpha_i^n \qquad (3.2)$$

*holds.*

(ii) *If furthermore* $e_1 = e_2 = \ldots = e_r = 1$ *and* D *denotes the discriminant of* f, *then for* $n=0,1,2,\ldots$ *one has*

$$u_n = \sum_{j=1}^{r} \mu_j \alpha_j^n$$

*with certain numbers* $\mu_1,\ldots,\mu_r \in K$ *which multiplied by* D *become integers of* K.

(iii) *Every sequence of the form* (3.2) *satisfies* (3.1) *with suitable* k *and* $a_0,a_1,\ldots,a_{k-1}$.

*Proof.* (i) Observe that the set of all solutions $\{u_n\}$ of (3.1) with complex terms is a linear space V over the complex numbers and since every such solution is fully determined by the finite sequence $u_0,\ldots,u_{k-1}$, which can be prescribed arbitrarily, V is k-dimensional. Now direct evaluation, utilizing the equality $f^{(s)}(\alpha_j) = 0$ $(j=1,2,\ldots,r;$ $s=0,1,\ldots,e_j)$ shows that for $j=1,2,\ldots,r;$ $i=0,1,\ldots,e_j-1$ the sequence

$$E_{ij} = \{n^i \alpha_j^n\}$$

forms a solution of (3.1). The set $\{E_{ij}\}$ is linearly independent, and since it has k elements it forms a basis of V. This implies the equality (3.2) with $P_i(x)$ being polynomials with complex coefficients. To show that in case when $u_n$ are rational numbers these coefficients lie already in K write

$$P_i(x) = \sum_{j=0}^{e_i-1} \beta_{ij}x^j \qquad (i=1,2,\ldots,r),$$

put here consecutively $x=0,1,\ldots,k-1$ and solve the system of linear equations resulting from (3.1). Since the $\alpha_i$'s are distinct, the determinant of this system is non-zero and lies in K. This proves $\beta_{ij} \in K$ and (i) results.

(ii) In the case $e_1 = \ldots = e_r = 1$ the determinant of the linear system, which was solved in the proof of (i) is a Vandermonde determinant, whose square equals D. Thus $D^{\frac{1}{2}}\beta_{ij}$ is for all choices of $i,j$ an algebraic integer and so is $D\beta_{ij}$, which obviously lies in K.

(iii) This assertion results from the observation that due to (3.2) the sequence $\{u_n\}$ is a linear combination of the $E_{ij}$'s. □

Note, that the integrality of the coefficients of $f(x)$ was used only in the proof of (ii).

2. Lemma 2.1 will now serve us to the proof of a necessary condition for UD(mod N) of a linear recurrent sequence in the case when the associated polynomial has no multiple roots.

THEOREM 3.2. *If* $\{u_n\}$ *is a linear recurrent sequence and its associated polynomial* $f(x)$ *has only simple roots, then* $\{u_n\}$ *can be* UD(mod p) *for a prime* p *only if* p *divides the discriminant of* f.

*Proof.* We use here elementary theory of ideals in the ring $Z_K$ of integers of the splitting field K of f. Let D be the discriminant of f and k its degree. It follows from part (ii) of the last lemma that with suitable integers $A_1,\ldots,A_k$ from the field K we have

$$Du = \sum_{j=1}^{r} A_j\alpha_j^n$$

for $n=1,2,\ldots,r$. Let p be a rational prime not dividing D and let

$$pZ_K = P_1^{t_1} \ldots P_s^{t_s}$$

be the factorization of the ideal generated by p in $Z_K$ into prime ideals. Observe now that for $j=1,2,\ldots,s$ and $i=1,2,\ldots,k$ the sequence $\alpha_i^n(\text{mod } P_j^{t_j})$ has its period $T_{ij}$ either equal to 1 (in case when

$\alpha_i \in P_j$) or to the order of $\alpha_i$ in the multiplicative group of residues (mod $P_j^{t_j}$) prime to $P_j$. Denoting by $T$ the least common multiple of the numbers $T_{ij}$ we infer from Lemma 3.1 (ii) that $T$ is a period of the sequence $\{u_n \pmod p\}$.

If now $\{u_n\}$ is UD(mod p) then in every its full period every residue class (mod p) will appear the same number of times, hence $T$ must be divisible by $p$. We shall now show that this leads to a contradiction. In the case when $K$ is the field of rational numbers, i.e. all roots $\alpha_1,\ldots,\alpha_k$ of $f$ are rational integers this is immediate, since in this case $s=1$, $t_1=1$, $P_1=pZ$ and since $T$ is a divisor of $\varphi(p)=p-1$ it cannot be a multiple of $p$. Assume now that $K \neq Q$. Denote by $\Phi_K(I)$ the *Euler's function* of the field $K$, i.e. the number of elements in the multiplicative group of residue classes (mod I), prime to $I$, for every non-zero ideal $I$ of $Z_K$. We need only two simple properties of this analogue of Euler's $\varphi$-function: its multiplicativity

$$\Phi_K(IJ) = \Phi_K(I)\Phi_K(J)$$

for coprime ideals $I$, $J$, and the explicit formula

$$\Phi_K(I) = N(I) \prod_{P|I} (1 - N(P)^{-1})$$

where the product is taken over all prime ideals dividing $I$, and $N(I)$ denotes the absolute norm of $I$, i.e. $N(I) = \# Z_K/I$ . Since for every $i,j$, $T_{ij}$ is a divisor of $\Phi_K(P_j^{t_j})$ we see that $T$ divides $\Phi_K(pZ_K)$. If now $N(P_j)=p^{f_j}$ $(j=1,2,\ldots,s)$, then $t_1f_1+\ldots+t_sf_s$ equals $[K:Q]$, the degree of $K/Q$ thus

$$\Phi_K(pZ_K) = p^{[K:Q]} \prod_{j=1}^{s} (1-p^{-f_j}) = p^E \cdot \prod_{j=1}^{s} (p^{f_j-1})$$

where

$$E = \sum_{j=1}^{s} (t_j-1)f_j \, .$$

It follows that $p$ can divide $\Phi_K(pZ_K)$ only if at least one of the integers $t_1,\ldots,t_s$ exceeds 1, in other words $p$ is ramified in $K/Q$ and according to the discriminant theorem (see e.g. W.NARKIEWICZ [74], cor.2 to th.4.8.) every such $p$ must divide the discriminant of

K, hence also the discriminant D of the polynomial f, but this contradicts our choice of p. □

COROLLARY. (L.KUIPERS, J.S.SHIUE [71], [72a]). *If the Fibonacci sequence is* UD(mod N) *then* N *must be a power of* 5.

*Proof.* The associated polynomial equals in this case $x^2 - x - 1$ and since its discriminant equals 5, the Fibonacci sequence is UD(mod p) only in case p=5. But if it is UD(mod N) then also it must be UD(mod p) for every prime p dividing N, hence the assertion. □

We shall show later (see the corollary to Theorem 3.5) that the converse to this corollary is also true. However the converse to the last theorem fails in general. To see this consider the *Lucas sequence* defined by $L_{n+2} = L + L_{n+1}$, $L_0 = 1$, $L_1 = 3$. Here the associated polynomial is the same as for the Fibonacci sequence, however this time there is no UD(mod 5) since the first terms of $L_n$ reduced (mod 5) are 1, 3, 4, 2, 1, 3 and so the period equals 4 and no term of our sequence is divisible by 5. Thus there is no integer with respect to which $L_n$ is uniformly distributed. This was first proved by L.KUIPERS, J.S.SHIUE [72b] who gave an elementary proof, without the use of algebraic number theory.

From Theorem 3.2 follows that if the polynomial f(x) is irreducible (mod p), then we cannot have UD(mod p). However, as shown by M.HALL [38a],[38b], in this case nevertheless certain uniformity of distribution occurs. In fact, if $T_p$ is the minimal period of $u_n$(mod p) and k is the order of $\{u_n\}$, then a given residue a(mod p) appears in the period $T_p/p + O(p^{k/2-1/2})$ times. (Of course this result is significative only in those cases when $T_p$ is comparatively large.)

## § 2. Uniform distribution (mod p) of second-order linear recurrences.

1. Necessary and sufficient conditions for a linear recurrent sequence of order k to be UD are known only in the case k=2,3 and 4. We shall treat here only the case k=2, since the conditions obtained in other cases are very cumbersome and one has the feeling that one should be able to find simpler conditions which, may be, would be generalizable to larger values of k.

We start with a criterion for UD(mod p) with a prime p.

THEOREM 3.3. (R.T.BUMBY [75], M.B.NATHANSON [75]). *Let* $u_n$ *be a recurrent sequence of second order with integral terms. Denote by* $f(x) = (x-a)(x-b) = x^2 - Ax - B$ *its associated polynomial and let* p *be a prime number.*

(i) *If* a,b *are rational then* $u_n$ *is* UD(mod p) *if and only if* p *divides the discriminant of* f *and does not divide* $B(u_1-au_0)$.

(ii) *If* a,b *are not rational, then* $u_n$ *is* UD(mod p) *if and only if* p *divides the discriminant of* f *and does not divide* $B(u_1^2 - Au_0u_1 - Bu_0^2) = ab(u_1 - au_0)(u_1 - bu_0)$

*Proof*. Assume first that $u_n$ is UD(mod p). Since in case a=b the discriminant D of f vanishes we get from Theorem 3.2 that in all cases p must divide D. Remembering that

$$u_{n+2} = Au_{n+1} + Bu \tag{3.3}$$

we see that if p divides B, then $u_{n+2} \equiv Au_{n+1} \pmod p$ and this implies $u_n \equiv u_0A^n \pmod p$ for n=1,2,3,... . If now either $u_0$ or A is divisible by p then $p|u_n$ for all $n \geq 1$ and otherwise $u_n$ is never divisible by p, so we cannot have UD(mod p).

Similarly, if in the case (i) $p|u_1-au_0$ then we get from (3.3) $u_n \equiv u_0a^n \pmod p$ obtaining a contradiction as before and if in the case (ii) $(u_1-au_0)(u_1-bu_0) = u_1^2 - Au_0u_1 - Bu_0^2$ is divisible by p, then with a proper choice of the prime ideal P dividing p in the ring $Z_K$ of integers of the field K(a) we get $u_1 \equiv au_0 \pmod P$ which in view of $a \equiv b \pmod P$ and $u_{n+2} \equiv 2au_{n+1} - a^2u_n \pmod P$ we obtain $u_n \equiv u_0a^n \pmod P$ for n=1,2,... which as before implies that either no $u_n$ lies in P and hence $p \nmid u_n$ or all $u_n$ $(n \geq 1)$ lie in P and since they are rational integers, they must be all divisible by p, ruling out UD(mod p).

2. Having proved the necessity of our conditions we turn to their sufficiency, and first dispose quickly of the case p=2. There are four quadratic polynomials over GF(2), viz. $x^2$, $x^2+1$, $x^2+x$ and $x^2+x+1$, hence f(x) must be congruent (mod 2) to one of them. The last two have discriminant 1, thus by the already proved part of the theorem we cannot expect UD(mod 2) in this case. If $f(x) \equiv x^2 \pmod 2$ then $u_{n+2}$ is even for $n \geq 2$ thus again UD(mod 2) is impossible. It remains the possibility $f(x) \equiv x^2 + 1 \pmod 2$, thus $u_{n+2} \equiv u_n \pmod 2$ and we shall

get UD(mod 2) if and only if $u_0$ and $u_1$ are of opposite parity. Now it is easy to check that the obtained conditions coincide with those stated in the theorem.

In the general case we shall base our argument on the following lemma:

LEMMA 3.4. *Let* $\alpha, \beta, \gamma, \delta$ *be integers in an algebraic number field* K, P *a prime ideal of* K *such that neither* $\beta$ *nor* $\delta$ *is divisible by it and let* $N(P) = p^t$. *Furthermore define*

$$f(x) = (\alpha + \beta x)\delta^h + \gamma,$$

*let* r *be the multiplicative order of* $\delta \pmod P$ (*i.e. the smallest positive integer with* $\delta^r \equiv 1 \pmod P$)) *and assume that for any* $s \geq 0$ *the unique solution* $x \bmod P$ *of the congruence*

$$(\alpha + \beta x)\delta^s + \gamma \equiv 0 \pmod P$$

*has a representative in* Z. *Then there are exactly* r *values of* $h \pmod{pr}$ *such that* $f(h) \equiv 0 \pmod P$.

*Proof.* For a fixed value of $\xi \pmod r$ the value $\delta^\xi \pmod P$ also becomes fixed. Our assumptions imply the existence of $\eta \in Z$ such that

$$(\alpha + \beta\eta)\delta^\xi + \gamma \equiv 0 \pmod P .$$

Since $\eta \pmod P$ is uniquely determined by $\xi$, the same can be asserted about $\eta \pmod p$, because $\eta$ is rational. As $r \mid N(P) - 1 = p^t - 1$ we get $(p,r) = 1$ and so the system

$$X \equiv \xi \pmod r , \qquad X \equiv \eta \pmod p$$

is satisfied by a rational integer $h = h(\xi)$, uniquely determined (mod pr). Clearly we have $f(h) \equiv 0 \pmod P$ and since $\xi \pmod r$ attains r values, the lemma follows. □

(It should be noted that the last condition is trivially satisfied when P is of the first degree, i.e. t=1, since in that case every residue class (mod P) contains rational integers.)

Having settled the case p=2 before assume now p to be odd. First we consider the case when f has a double root, i.e. $f(x) = (x-a)^2$

with $a \in Z$ and $p \nmid a(u_1 - au_0)$. Lemma 3.1 (i) shows that in this case

$$u_n = (\beta_0 + \beta_1 n)a^n \tag{3.4}$$

holds for $n \geq 0$ with suitable rational $\beta_0, \beta_1$. It follows that $\beta_0 = u_0$ and $\beta_1 = u_1/a - u_0$ thus denoting by $\bar{a}$ an integer satisfying $a\bar{a} \equiv \equiv 1 \pmod p$ and putting $M_0 = \beta_0$, $M_1 = u_1\bar{a} - u_0$ we get

$$u_n \equiv (M_0 + M_1 n)a^n \pmod p .$$

If now $r$ denotes the multiplicative order of $a \pmod p$ then the last congruence shows that $pr$ is one of the periods of $u_n \pmod p$. It remains to show that among every $pr$ consecutive elements of this sequence every residue class $\pmod p$ will appear exactly $r$ times and for this purpose we utilize Lemma 3.4. In fact, for any integer $h$ we have

$$u_{n+h} - u_n \equiv \{(M_0 + M_1 n + M_1 h)a^h - (M_0 + M_1 n)\}a^n \pmod p$$

applying that lemma with $\delta = a$, $\alpha = M_0 + M_1 n$, $\beta = M_1$, $\gamma = -\alpha$, $P = p$ (which is legitimate, since if $p$ would divide $M_1$ then it would also divide $u_1 - au_0$, contrary to our assumptions) we obtain that every residue class $\pmod p$ attained by the sequence $u_n$ occurs exactly $r$ times among consecutive $pr$ terms, and this implies that every residue class $\pmod p$ is actually attained with the right frequency.

Now assume that $f$ has two simple roots, i.e. $f(x) = (x-a)(x-b)$ with $a,b$ being distinct integers of $K$, the splitting field of $f$, which is either equal to the rational field $Q$ or to its quadratic extension. Our assumptions in this case may be restated as follows: if $P$ is a prime ideal dividing $p$ then $a \equiv b \pmod P$ and $ab(u_1 - au_0) \not\equiv \not\equiv 0 \pmod P$. (Of course in case $K=Q$ we have $P=p$). Lemma 3.1 (ii) shows that with suitable $\beta_1, \beta_2 \in K$ we have

$$u_n = \beta_1 a^n + \beta_2 b^n \qquad (n \geq 0) .$$

Putting here $n=0,1$ and solving the resulting system for $\beta_1, \beta_2$ we get $\beta_1 = (u_1 - u_0 b)/(a-b)$, $\beta_2 = (u_0 a - u_1)/(a-b)$ thus

$$u_n = \frac{(u_1 - u_0 b)a^n + (u_0 a - u_1)b^n}{a-b} = u_1 \frac{a^n - b^n}{a-b} - u_0 ab \frac{a^{n-1} - b^{n-1}}{a-b} \tag{3.5}$$

Due to $a \equiv b \pmod{P}$ this may be simplified in view of

$$\frac{a^t - b^t}{a - b} = \sum_{j=0}^{t-1} a^j b^{t-j} \equiv t a^{t-1} \pmod{P} \qquad (t \geq 1)$$

to the handable form

$$u_n \equiv n u_1 a^{n-1} - (n-1) u_o a^n \pmod{P}$$

thus

$$u_n \equiv a^{n-1}(n u_1 - (n-1) u_o a) \pmod{P} .$$

and we see that $u \pmod{P}$ has $pr$ for its period, with $r$ being the multiplicative order of $a \pmod{P}$. The remainder of the argument imitates that of the previous case: we have for $h \geq 1$

$$u_{n+h} - u_n \equiv a^{n+h-1}((n+h)u_1 - (n+h-1)u_o a) - a^{n-1}(n u_1 - (n-1)u_o a) \equiv$$

$$\equiv a^{n-1}((\alpha + \beta h) a^h - \alpha) \pmod{P}$$

with $\alpha = n u_1 - (n-1) u_o a$, $\beta = u_1 - u_o a$ .

Now we have to distinguish between two possibilities:

I. $N(P) = p$. In this case Lemma 3.4 is directly applicable and implies that among $pr$ consecutive elements of $\{u_n\}$ every residue class $(\text{mod } P)$ which is attained by this sequence occurs $r$ times. Since $N(P) = p$ we infer that every residue class $(\text{mod } P)$ is attainable and so every residue class $(\text{mod } p)$ appears in the sequence $\{u_n\}$ with the same frequency, proving uniform distribution $(\text{mod } p)$.

II. $N(P) = p^2$. (Hence $K \neq Q$ and $p = P$ thus $p$ is unramified in $K$.) Here we shall also apply Lemma 3.4 however first we have to ascertain that its assumptions are satisfied. (This is the only place in the proof, where the assumption $p \neq 2$ will be used.) If $f(x) = x^2 - Ax - B$, $D$ is the discriminant of $f$ and $d$ the discriminant of $K$, then in view of $p | D$, $p \neq 2$, $p \nmid d$ (because all prime ideals dividing the discriminant of a quadratic number field are of the first degree) we must have $p^2 | D$ and so we may write $D = p^2 R^2 d$ with a certain rational integer $R$. Thus $a = (A + pR\sqrt{d})/2$ and with $C = (p+1)/2$ we get

$$a \equiv (A + pR\sqrt{d}) C \equiv AC \pmod{P}$$

Putting for shortness $T = nu_1 - (n-1)u_0C$, $U = u_1 - u_0C$ we obtain now

$$\alpha \equiv T(\mathrm{mod}\ P)\ , \qquad \beta \equiv U(\mathrm{mod}\ P)$$

which gives

$$(\alpha + \beta h)a^s - \alpha \equiv (T + Uh)(AC)^s - T(\mathrm{mod}\ p)$$

Since $T,U,A,C$ are rational integers one may find $h \in Z$ such that $(\alpha+\beta h)a^s - \alpha \equiv 0(\mathrm{mod}\ p)$ and this shows that the conditions of Lemma 3.4 are satisfied in our case. Hence in every sequence of $pr$ consecutive elements of $\{u_n\}$ every attainable residue class (mod P) appears $r$ times. Now only those residue classes can be attained which have a representative in $Z$ and since there are $p$ such classes we obtain that every residue class (mod p) occurs in the sequence $\{u_n\}$ with the same frequency, and UD(mod p) follows. □

3. Note that H.NIEDERREITER, J.S.SHIUE [77] characterized second-order linear recurrences which are UD(mod p) in the following, equivalent form: $\{u_n\}$ *is* UD(mod p) *if and only if its associated polynomial has a multiple root* (mod p) *and the sequence* $v_n = u_n(\mathrm{mod}\ p)$ *does not satisfy a recurrence relation of order lower than* 2.

To apply this criterion it is necessary to have a way of checking, whether a given recurrent sequence $u_n(\mathrm{mod}\ p)$ does satisfy a recurrence relation of lower order. Here an old result of L.KRONECKER [81] (§ VII) is useful, which states that this happens if and only if the determinant of the matrix

$$\begin{bmatrix} u_0 & u_1 & \cdots & u_{k-1} \\ u_1 & u_2 & \cdots & u_k \\ \cdots & \cdots & \cdots & \cdots \\ u_{k-1} & u_k & \cdots & u_{2k-2} \end{bmatrix}$$

vanishes (mod p). For another proof of this see M.WILLETT [76].

The paper of H.NIEDERREITER, J.S.SHIUE [77] contains also the following characterization of recurrences of third degree, which are UD(mod p):

Assume that $u_n$ is a third order recurrent sequence, such that the sequence $u_n(\mathrm{mod}\ p)$ is not a recurrent sequence of order $\leq 2$ and its associated polynomial $f$ satisfies

$$f(x) \equiv (x-a)^2(x-b) \pmod p$$

with $p \nmid ab$, then $u_n$ is UD(mod p) if and only if either $a \not\equiv b \pmod p$ or $a \equiv b \pmod p$ and $p=2$ or finally, $a \equiv b \pmod p$, $p \neq 2$, a is a quadratic non-residue (mod p) and the congruence

$$(u_2 - 4au_1 + a^2u_0)^2 \equiv 4a^2u_0u_2 \pmod p$$

holds. (Cf. also M.J.KNIGHT, W.A.WEBB [80]).

A similar characterization of uniformly distributed recurrent sequences of order four is stated in the same paper, and its proof appears in H.NIEDERREITER, J.S.SHIUE [80]. The conditions become extremely complicated and so we refrain from quoting them here.

## § 3. General modulus

1. Having settled the case of a prime modulus we turn now to the general case: (R.T.BUMBY [75]; W.A.WEBB, C.T.LONG [75] for $N = p^k$).

THEOREM 3.5. *Let* $\{u_n\}$ *be a recurrence sequence of order two, whose terms are rational integers let*

$$f(x) = x^2 - Ax - B$$

*be its associated polynomial, and* $D = A^2 + 4B$ *its discriminant.*

*If* N *is a given integer, then* $\{u_n\}$ *is* UD(mod N) *if and only if* $\{u_n\}$ *is* UD(mod p) *for all primes* p *dividing* N, *provided none of the following exceptional cases holds:*

(*a*) $4 \mid N$, $A \not\equiv 2 \pmod 4$,

(*b*) $4 \mid N$, $B \not\equiv 3 \pmod 4$,

(*c*) $9 \mid N$, $A^2 + B \equiv 0 \pmod 9$.

*In these three cases* $\{u_n\}$ *is not uniformly distributed* (mod N).

*Proof.* First we shall prove that in the cases (a), (b), (c) one cannot expect uniform distribution.

In the case (a) we may assume N=4. If $f(x) = (x-a)^2$ then using Theorem 3.3 (i) we get $2 \nmid a$ and thus $A = -2a \equiv 2 \pmod 4$ so that (a) in fact never occurs in that case. If f has distinct roots, then again by Theorem 3.3 we get $2|D=A^2+4B$ thus $A$ is even and $A\not\equiv 2 \pmod 4$ implies $4|A$. Thus $u_{n+2}\equiv Bu_n \pmod 4$ and if for a certain $n$ we have $u_n\equiv 0 \pmod 4$ then $0\equiv u_{n+2j} \pmod 4$ holds for all $j\ge 0$ and we cannot have uniform distribution (mod 4).

In case (b) we again assume $N=4$ and since $A$ is even we can, due to (a), assume that $A\equiv 2 \pmod 4$. If $B$ were even then $u_n$ would be even for $n\ge 3$ so we may also assume $B\equiv 1 \pmod 4$ thus $u_{n+2}\equiv$ $\equiv 2u_{n+1}+u_n \pmod 4$. If $u_0, u_1$ are of the same parity, then all terms $u_n$ have the same parity, and so there is no UD(mod 4) in this case. If however $u_0\not\equiv u_1 \pmod 2$ then either $u_0\equiv 2u_1+u_o \pmod 4$ (in case $2|u_1$) or $u_1\equiv 2u_0+u_1 \pmod 4$ (in case $2|u_0$) and since the first terms of the sequence $\bar{u}_n=u_n \pmod 4$ are equal $\bar{u}_0, \bar{u}_1, 2\bar{u}_1+\bar{u}_0, 2\bar{u}_0+\bar{u}_1, \bar{u}_0, \bar{u}_1, \ldots$ this sequence is of period four and in this period one of the residue classes (mod 4) appears twice making UD(mod 4) impossible.

In case (c) we assume $N=9$ and $A^2+B\equiv 0 \pmod 9$. If both $A$ and $B$ are divisible by 3 then $3|u_n$ for $n\ge 2$, so assume this not to be the case. This leaves us with the following six possibilities for the pairs [A mod 9, B mod 9]: [1,8], [2,5], [4,2], [5,2], [7,5], [8,8]. Proceeding as in the preceding case one obtains with a little patience that the period of $u_n \pmod 9$ is in these cases either three or six, ruling out the possibility of uniform distribution (mod 9), since in that case the period should be a multiple of 9. We leave the full details to the reader, and consider here only the worst case, viz. $A\equiv 4 \pmod 9$, $B\equiv 2 \pmod 9$ in which a rather long preperiod occurs. Writing for shortness $x=u_o \pmod 9$, $y=u_1 \pmod 9$ we compute the first 14 terms of $u_n \pmod 9$: $x$, $y$, $2x+4y$, $7y$, $4x$, $7x+5y$, $2y$, $5x$, $2x+4y$, $3x+6y$, $7x+5y$, $7x+5y$, $6x+3y$, $2x+4y$, $2x+4y$, $3x+6y$, thus after a preperiod of eight terms we get a period of length six.

This proves the "only if" part of the theorem.

2. In the proof of the "if" part we shall use two easy auxiliary results, which will be now established:

LEMMA 3.6. *Let* $\{a_n\}$ *be a sequence of integers,* $N_1$ *a given integer and* $p$ *a prime, not dividing* $N_1$. *Let also* $\alpha$ *be a non-negative integer and assume that the sequence* $\{a_n\}$ *is* UD(mod $N_1p^\alpha$). *If there exists a positive integer* $h$ *with the property, that for* $j=0,1,\ldots,h-1$ *the sequence*

$$b_n = b_n^{(j)} = a_{j+hn} \qquad (n=1,2,3,\ldots)$$

*is constant* $(\bmod N_1p^{\alpha})$ *for all large* $n$, *say*

$$b_n^{(j)} \equiv \gamma(j) (\bmod N_1p^{\alpha})$$

*and moreover there is an integer* $c = c(j)$ *not divisible by* $p^{1+\alpha}$ *such that*

$$b_{n+1} - b_n \equiv c (\bmod p^{1+\alpha})$$

*holds for all large* $n$, *then the sequence* $\{a_n\}$ *is* $UD(\bmod N_1p^{1+\alpha})$

*Proof*. We may certainly assume that all asumptions hold for $n \geq 0$. Now fix $j \in [0,h-1]$. The sequence $b_n^{(j)} (\bmod p^{1+\alpha})$ is periodic and since $p^{\alpha} \| c(j)$ we obtain that $p$ is a period of it. Obviously every residue class $(\bmod p^{1+\alpha})$, congruent to $\gamma(j) (\bmod p^{\alpha})$ occurs exactly once in that period. If follows that if $\gamma$ is a residue class $(\bmod p^{1+\alpha})$ which satisfies $\gamma \equiv \gamma(j) (\bmod p^{\alpha})$, then

$$\# \{ n \leq x: b_n \equiv \gamma (\bmod p^{1+\alpha}) \} = x/p + o(x)$$

and if $\beta$ is a residue class $(\bmod N_1p^{1+\alpha})$ satisfying $\beta \equiv \gamma(j) (\bmod N_1p^{1+\alpha})$, then

$$\# \{ n \leq x: b_n \equiv \beta (\bmod N_1p^{1+\alpha}) \} = x/p + o(x)$$

Observe now that since the sequence $\{a_n\}$ is $UD(\bmod N_1p^{\alpha})$ it follows that the number of indices $j$ such that $\gamma(j)$ lies in a given residue class $(\bmod N_1p^{\alpha})$ equals $h/N_1p^{\alpha}$ and thus for every $\beta$

$$\#\{n \leq x: a_n \equiv \beta (\bmod N_1p^{1+\alpha}) \} = \sum_j \#\{m \leq \frac{x-j}{h}: b_m^{(j)} \equiv \beta (\bmod N_1p^{1+\alpha}) \}$$

(where the sum is taken over those $j$'s for which $\gamma(j) \equiv \beta (\bmod N_1p^{\alpha})$), and this equals further

$$\sum_j (x/ph + o(x)) = hx/N_1p^{1+\alpha}h + o(x) = x/N_1p^{1+\alpha} + o(x)$$

which shows that our sequence is $UD(\bmod N_1p^{1+\alpha})$. □

LEMMA 3.7. *If* $K$ *is an algebraic number field,* $x$ *an integer of* $K$*,* $P$ *a prime ideal of* $Z_K$*, containing the prime number* $p$*,* $x \equiv 1 \pmod{P}$ *and* $s$ *is defined by* $P^s \| x-1$ *then*

$$P^{s+w} | x^p - 1 \qquad \text{if} \quad sp \geq s+w ,$$

$$P^{s+w} \| x^p - 1 \qquad \text{if} \quad sp > s+w ,$$

*where* $w$ *is defined by* $P^w \| p$.

*Proof.* If $y = x-1$, then $y \in P^s \setminus P^{s+1}$ and $x^p = (1+y)^p = 1 + py + cpy^2 + y^p \equiv$ $\equiv 1 + py + y^p \pmod{P^{2s+w}}$ holds with an integer $c$, and in view of $P^{s+w} \| py$, $P^{sp} \| y^p$ our assertion follows.

COROLLARY 1. *If* $x, y$ *are integers of* $K$ *and* $P$ *a prime ideal of* $Z_K$ *such that with a positive* $s$ *one has* $P^s \| x-y$ *and* $P \nmid xy$*, then*

$$P^{s+w} | x^p - y^p \qquad \text{if} \quad sp \geq s+w ,$$

$$P^{s+w} \| x^p - y^p \qquad \text{if} \quad sp > s+w ,$$

*where* $w$ *has the same meaning as in the lemma.*

*Proof.* Let $z$ be an integer of $K$ satisfying $yz \equiv x \pmod{P^{s+w+1}}$. Then $x^p - y^p \equiv y^p(z^p - 1) \pmod{P^{s+w+1}}$ and $x-y \equiv y(z-1) \pmod{P^{s+w+1}}$ thus it suffices to apply the lemma to the number $z$. □

COROLLARY 2. *Under the assumptions of corollary 1 one has for* $j = 1, 2, \ldots$

$$P^{s+jw} | x^{p^j} - y^{p^j} \qquad \text{if} \quad sp \geq s+w ,$$

$$P^{s+jw} \| x^{p^j} - y^{p^j} \qquad \text{if} \quad sp > s+w .$$

*Proof.* Follows by induction from the previous corollary. □

3. Now we turn to the proof of sufficiency of the conditions stated in the theorem. It is enough to prove the following proposition, since then the theorem will follow by induction in the number of prime factors of the modulus, counted according to their multiplicity:

PROPOSITION 3.8. *Let* $\{u_n\}$ *be a sequence satisfying the conditions of Theorem 3.5. Let* N *be an integer and* p *a prime. Assume that* p *is at least equal to the largest prime divisor of* N *and assume further that none of the following three conditions holds:*

(*a*) $p=2$, $N=2^u$ $(u\geq 1)$, $A\not\equiv 2 \pmod 4$,

(*b*) $p=2$, $N=2^u$ $(u\geq 1)$, $B\not\equiv 3 \pmod 4$,

(*c*) $p=3$, $N=2^u 3^v$ $(u\geq 0,\ v\geq 1)$, $A^2+B\equiv 0 \pmod 9$.

*Then, if* $\{u_n\}$ *is* UD(mod p) *and* UD(mod N) *then it is also* UD(mod pN).

*Proof*. First we shall treat the easier case, when the associated polynomial has a double root, i.e. $f(x)=(x-a)^2$ with $a\in\mathbb{Z}$. Write $N=N_1p^k$ with $p\nmid N_1$ and $k\geq 0$. From (3.4) we get

$$u_n = (u_0+(u_1a^{-1}-u_0)n)a^n \qquad (n=0,1,\ldots)$$

and since by Theorem 3.3 (i) we must have $(a,N_1p)=1$ we may replace $u_n$ by $v_n=au_n$, without spoiling the UD-property. Thus

$$v_n = (au_0+(u_1-au_0)n)a^n \qquad (n=0,1,\ldots)$$

and for every $h\geq 0$ we get

$$v_{n+h}-v_n = a^n((c_1+c_2h)a^h-c_1)$$

with $c_2=u_1-au_0$ and $c_1=au_0+c_2n$. Put now

$$h = p^k(p-1)N_1\varphi(N_1) = N_1\varphi(p^{1+k}N_1) .$$

Then $p^kN_1 | h$ and $a^h\equiv 1 \pmod{p^{1+k}N_1}$ thus

$$v_{n+h}-v_n \equiv 0 \pmod{p^kN_1} .$$

If we would have also

$$v_{n+h}-v_n \equiv 0 \pmod{p^{1+k}N_1} ,$$

then

$$0 \equiv (c_1+c_2h)a^h-c_1 \equiv c_2h \pmod{p^{1+k}N_1}$$

however Theorem 3.3 implies $p \nmid c_2$ and if $p^{1+k} | h$, then $p | N_1 \varphi(N_1)$, however $p \nmid N_1$ and if $q$ is a prime dividing $N_1$, then $q < p$, hence $p \nmid q(q-1)$ and thus $p$ cannot divide $\varphi(N_1)$. But it cannot divide $N_1$ neither and so we see that $h$ is not divisible by $p^{1+k}$.

Because for any $j$ the value of $a^n (\mathrm{mod}\ p^{1+k}N_1)$ is constant for $n \equiv j (\mathrm{mod}\ h)$ it follows that the sequence $\{u_n\}$ satisfies the assumptions of Lemma 3.6 and applying it we obtain $UD(\mathrm{mod}\ p^{1+k}N_1)$ for $u_n$.

4. Now let us assume that the associated polynomial $f(x) = x^2 - Ax - B = (x-a)(x-b)$ has simple roots. In this case we utilize the formula (3.5) which readily implies

$$u_{n+h} = a^h(\beta_1 a + \beta_2 b^n) + \beta_2 b^n (a^n - b^n) = a^h u_n + (u_0 a + u_1) b^n \frac{a^h - b^h}{a - b}$$

for all $h \geq 0$, hence

$$u_{n+h} - u_n = (a^h - 1) u_n + c b^n \frac{a^h - b^h}{a - b} \tag{3.7}$$

with $c = u_0 a - u_1$.

Write again $N = N_1 p^k$ with $p \nmid N_1$, $k \geq 0$, let $K$ be the splitting field of $f$ (hence $K$ equals either $Q$ or a quadratic extension of it), $P$ prime ideal of $Z_K$ lying over $p$, define $w$ by $P^w \| p$ and for any prime ideal $q$ define $\lambda_q$ by $q^{\lambda_q} \| a-b$. This notation will be utilized in the next sequence of lemmas.

Observe that due to Lemma 3.6 and (3.7) the proposition will follow once we find an integer $h_k$ satisfying the following three conditions:

(i) $$\frac{a^{h_k} - b^{h_k}}{a - b} \equiv 0 (\mathrm{mod}\ N_1 p^k),$$

(ii) $$\frac{a^{h_k} - b^{h_k}}{a - b} \not\equiv 0 (\mathrm{mod}\ p^k P),$$

and

(iii) $$a^{h_k} \equiv b^{h_k} \equiv 1 (\mathrm{mod}\ N_1 p^k P).$$

Indeed, (i) and (iii) imply that for $h = h_k$, $j = 0, 1, \ldots, h-1$ and $b_m = u_{j+hm}$ the congruence $b_{m+1} - b_m \equiv 0 (\mathrm{mod}\ N_1 p^k)$ holds, i.e. $b_m (\mathrm{mod}\ N_1 p^k)$

is constant. Further, since for $n = j+hm$ we have $b^n \equiv b^j \pmod{p^k P}$ and $a^h - 1 \equiv 0 \pmod{p^k P}$ (due to (iii)) it follows from (3.7) that the difference $b_{m+1} - b_m$ is congruent $(\bmod\ p^k P)$ to

$$cb^j \frac{a^h - b^h}{a - b},$$

which is a constant, not congruent to zero by (ii). Since $b_{m+1} - b_m$ is a rational integer, the same assertion holds also $(\bmod\ p^{1+k})$ and thus the Lemma 3.6 is applicable.

Before we prove the existence of $h_k$ observe that in the case $p=2$ we can restrict ourselves to $k \geq 1$, since for $k=0$ we get $N_1 = 1$ and the assertion becomes obvious.

First we reduce the problem to the case $k=0$ resp. $k=1$:

LEMMA 3.9. *If* $r \geq 1$ *and* $h_r$ *is an integer satisfying* (*i*) - (*iii*) *with* $k = r$, *then* $h_{r+1} = ph_r$ *satisfies* (*i*) - (*iii*) *for* $k = r+1$. *The same holds for* $r = 0$, *except when either* $p = 2$, *or* $p = 3$, $w = 2$, $\lambda_P = 1$.

*Proof*. The truth of (i) and (iii) for $p \neq 2$ follows immediately from the Corollary 1 to Lemma 3.7. The same holds also for (ii), with the exception however of the case $p = 3$, $w = 2$, $\lambda_P = 1$, the only case where the assumption $sp > s+w$ is violated (with $s = \lambda_P$). Indeed, if $sp \leq s+w$, then $s(p-1) \leq w$ and since $w \leq 2$, $s \geq 1$ and $p \geq 2$ it follows that either $p = 2$ or $p = 3$, $s = 1$, $w = 2$. □

LEMMA 3.10. *If* $p \neq 2$ *then with the exception of the case when* $p = 3$, *and 2 generates a prime ideal the conditions* (*i*) - (*iii*) *are satisfied for* $k = 0$ *with*

$$h_0 = \Phi_K\Big(PN_1 \prod_{q \mid N_1} q^{\lambda_q}\Big)$$

*where the product is taken over all prime ideals* $q$ *dividing* $N_1$.

*Proof*. The conditions (i) and (iii) are clearly satisfied, since by Theorem 3.3 we have $(ab, PN_1) = 1$ and thus

$$a^{h_0} \equiv b^{h_0} \equiv 1 \Big(\bmod\ PN_1 \prod_{q \mid N_1} q^{\lambda_q}\Big).$$

Moreover by the same theorem we have $a \equiv b \pmod{P}$, hence

$$\frac{a^{h_o} - b^{h_o}}{a - b} = \sum_{j=0}^{h_o-1} a^j b^{h_o-j-1} \equiv \sum_{j=0}^{h_o-1} a^{h_o-1} \equiv h_o a^{h_o-1} \pmod{P} \tag{3.8}$$

and to show that $h_o \not\equiv 0 \pmod{P}$ observe that any prime divisor of $h_o$ either divides $p(p-1)(p+1)$ or $p_1(p_1-1)(p_1+1)$ for a certain prime $q_1 | N_1$. In fact, if for prime ideals $q$ dividing $N_1$ we define $\alpha_q$ by $q^{\alpha_q} \| N_1$, then

$$h_o = (N(P) - 1) \prod_{q | N_1} N(q)^{\alpha_q + \alpha_q - 1} (N(q) - 1)$$

and $N(P) = p$ or $p^2$, $N(q) = p_1$ or $p_1^2$ for a suitable $p_1 | N_1$. Since all $p_1$'s are smaller than $p$, the only possibility of $p | h_o$ arises in case when for certain $q$, $p_1$ we have $p | N(q) - 1 = (p_1-1)(p_1+1)$, thus $p | p_1+1$ and since $p_1 < p$ this is possible only if $p = 3$, $p_1 = 2$ and 2 generates a prime ideal $q$ in $Z_K$, but this case we did exclude. Thus $p \nmid h_o$. □

5. From these lemmas we obtain our proposition for all $p \geq 3$, except when $p = 3$ and either 2 remains prime in $K$ or $w = 2$ and $\lambda_p = 1$. Now we have to deal with the remaining cases. First let $p = 3$.

LEMMA 3.11. *If* $p = 3$ *then with* $h_o = N_1$ *the conditions (i) - (iii) will be satisfied for* $r = 0$. *If* $p = 3$, $w = 2$, $\lambda_p = 1$, *and* $A^2 + B \not\equiv 0 \pmod{a}$ *then (i) - (iii) will be satisfied with* $h_1 = 3N_1$ *for* $r = 1$.

*Proof.* We start with the case $r = 0$. Here $N_1 = 2^m$ with a suitable $m \geq 1$ since for $m = 0$ the proposition is evident. The conditions (i) - (iii) get the follwoing form:

$$\frac{a^{2^m} - b^{2^m}}{a - b} \equiv 0 \pmod{2^m}, \tag{3.9}$$

$$\frac{a^{2^m} - b^{2^m}}{a - b} \not\equiv 0 \pmod{P} \tag{3.10}$$

and

$$a^{2^m} \equiv b^{2^m} \equiv 1 \pmod{2^m P}. \tag{3.11}$$

Let $P_2$ be a prime ideal dividing 2, let $P_2^t \| 2$ and define $u$ by $P_2^u \| a-b$. Since $(a^2-b^2)/(a-b) = a+b \equiv a-b \equiv 0 \pmod 2$ we get $a^2-b^2 \equiv$ $\equiv 0 \pmod{P_2^{u+t}}$ and it follows from Corollary 1 to Lemma 3.7 that for all $m \geq 1$ we have

$$\frac{a^{2^m}-b^{2^m}}{a-b} \equiv 0 \pmod{P_2^{mt}} \equiv 0 \pmod{2^m}$$

showing that the first condition holds.

To check the second condition note that since $a \equiv b \pmod P$ we get

$$\frac{a^{2^m}-b^{2^m}}{a-b} = \sum_{j=0}^{2^m-1} a^j b^{2^m-j-1} \equiv 2^m a^{2^m-1} \not\equiv 0 \pmod P)$$

For the third condition one has to recall, that since $u_n$ is assumed to be UD(mod 2) we have $f(x) \equiv x^2-1 \pmod 2$ thus $a^2 \equiv b^2 \equiv 1 \pmod 2$ and the application of the Corollary 2 to Lemma 3.7 in the case of prime ideal over 2 leads to

$$a^{2^m} \equiv b^{2^m} \equiv 1 \pmod{2^m}.$$

Since the polynomial $f(x)$ mod 3 has $a \pmod P$ for its double root, a must be congruent (mod P) to a rational integer, i.e. a(mod P) lies in GF(3), so $a^2 \equiv 1 \pmod P$ must hold and we obtain immediately $a^{2^m} \equiv b^{2^m} \equiv 1 \pmod P$.

This establishes the lemma in the case $r = 0$. In the case $r = 1$ we have to assume $w = 2$, $\lambda_P = 1$ and since $N_1 = 2^m$ $(m \geq 0)$, the conditions (i) - (iii) take the following shape:

$$\frac{a^{3\cdot 2^m}-b^{3\cdot 2^m}}{a-b} \equiv 0 \pmod{3\cdot 2^m}$$

$$\frac{a^{3\cdot 2^m}-b^{3\cdot 2^m}}{a-b} \not\equiv 0 \pmod{3\cdot 2^m\cdot P}$$

$$a^{3\cdot 2^m} \equiv b^{3\cdot 2^m} \equiv 1 \pmod{3\cdot 2^m\cdot P} .$$

If $m \geq 1$ then the first and third conditions are consequences of (3.9) resp. (3.10) by Corollary 1 to Lemma 3.7.

In case $m=0$ the same argument applies, this time based on $\lambda_P=1$, i.e. $P\|a-b$. Thus we are left with the second condition, which it suffices to consider in the case $m=0$. If it is violated, then $3P|(a^3-b^3)/(a-b)=3a^2+3a(b-a)+(a-b)^2$ hence $3a^2+(b-a)^2\equiv 0 \pmod{3P}$. We shall show that this leads to the excluded case $A^2+B\equiv 0 \pmod 9$.

In fact, writing $a=(A+D^{\frac{1}{2}})/2$, $b=(A-D^{\frac{1}{2}})/2$ we obtain that the last congruence is equivalent to

$$A+B+6AD^{\frac{1}{2}} \equiv 0 \pmod{3P}$$

however $6AD^{\frac{1}{2}}\equiv 0 \pmod{3P}$, thus $A^2+B\equiv 0 \pmod{3P}$ and since $A^2+B$ is a rational integer, the same congruence $(\mathrm{mod}\ 3^2)$ results. □

It remains the case $p=2$, in which clearly $N_1=1$.

LEMMA 3.12. *In the case* $p=2$ *the conditions (i) - (iii) will be satisfied in case* $r=1$ *by* $h_1=2$, *except when either* $A\not\equiv 2 \pmod 4$ *or* $B\not\equiv 3 \pmod 4$.

*Proof*. Since $(a^2-b^2)/(a-b)=a+b=A\equiv 2 \pmod 4$ the conditions (i) and (ii) hold, and to show that $a^2\equiv b^2\equiv 1 \pmod{2P}$ observe that if $A=4A_1+2$, then

$$a^2=(A^2+D+2AD^{\frac{1}{2}})/4=8A_1^2+8A_1+2+B+2A_1D^{\frac{1}{2}}+D^{\frac{1}{2}}\equiv 2+B+2A_1D^{\frac{1}{2}}+D^{\frac{1}{2}} \pmod 4$$

where $D=A^2+4B$. But $D=4(4(A_1^2+A_1)+1+B)$ is in view of $B\equiv 3 \pmod 4$ divisible by 16, thus $D^{\frac{1}{2}}$ is divisible by 4, so finally

$$a^2\equiv 2+B\equiv 1 \pmod 4$$

and the assertion about $b$ follows by conjugation in case $K\neq Q$ and is anyway obvious if $K=Q$. □

The proposition results now immediately, and, as already noted, so does the theorem. □

6. To give an application, we consider the Fibonacci sequence and its relatives, defined by $u_{n+2}=u_{n+1}+u_n$ with prescribed $u_0$ and $u_1$. (In the Fibonacci case $u_0=u_1=1$.) We saw already (in the Corollary to Theorem 3.2) that the Fibonacci sequence can be $\mathrm{UD}(\mathrm{mod}\ N)$ only if $N$ is a power of 5 and the same applies also to sequences with another

choice of the initial elements. The last theorem shows now that $\{u_n\}$ will be $UD(\mathrm{mod}\ 5^k)$ if and only if it is $UD(\mathrm{mod}\ 5)$. Since the roots of the associated polynomial are

$$a = (1+5^{1/2})/2\ , \qquad b = (1-5^{1/2})/2$$

and $5\nmid ab$, we obtain from Theorem 3.3 (ii) that $\{u_n\}$ will be $UD(\mathrm{mod}\ 5)$ if and only if

$$5\nmid N(u_1 - au_0) = ((2u_1-u_0)^2 - 5u_0^2)/4 = u_1^2 - u_0u_1 - u_0^2\ .$$

If $5|u_0$, then this condition will be satisfied for any $u_1$ not divisible by 5 and if $5\nmid u_0$, then putting $u_1 \equiv tu_0 (\mathrm{mod}\ 5)$ we can transform this condition to

$$0 \not\equiv u_0^2(t^2-t-1) \equiv u_0^2(t-3)^2 (\mathrm{mod}\ 5)$$

and one sees that it holds if and only if $5\nmid u_0$, and $u_1 \not\equiv 3u_0 (\mathrm{mod}\ 5)$. Finally we can claim the following

COROLLARY. (L.KUIPERS, J.S.SHIUE [71]). *If* $u_{n+2} = u_n + u_{n+1}$ *then the sequence* $\{u_n\}$ *is* $UD(\mathrm{mod}\ N)$ *if and only if* $N$ *is a power of* 5 *and* $5\nmid u_1 - 3u_0$. □

In particular, the Fibonacci sequence is $UD(\mathrm{mod}\ N)$ if and only if $N$ is a power of 5. The sufficiency of this condition was established by H.NIEDERREITER [72a] and the necessity by L.KUIPERS and J.S.SHUIE [71], [72a]. Another proof gave P.BUNDSCHUH [74]. Note also, that the results of A.P.SHAH [62], G.BRUCKNER [70] imply that for no prime $p \geq 11$ can the Fibonacci sequence contain all residues (mod p).

## § 4. Notes and comments

1. The first general study of linear recurrences was done by E.LUCAS [78] and the first results which are connected with uniform distribution can be found in papers of R.D.CARMICHAEL [20], H.T.ENGSTROM [31] and M.WARD [31a], [31b], [33] where the problem of determining the least

period $v(p)$ of $u_n \pmod N$ (with $u_n$ being a given linearly recurrent sequence) was considered. Carmichael proved a.o. that for prime powers $p^t$ one has $v(p^t) = p^b v(p)$ with a certain $b \in [0,t]$ and M.Ward in the first quoted paper showed that $b \le t-1$. His second paper is devoted to recurrences of the third order and contains a wealth of results concerning the distribution of residues $\pmod p$ of such a recurrence. Engstrom's paper applies algebraic number theory to the study of the general period $\pmod p$ of a recurrence, i.e. a number which is a period $\pmod p$ for any recurrence with the same associated polynomial. From newer papers concerning the periods $\pmod N$ of recurrences cf. A.VINCE [81], who utilized matrix theory to determine them.

The knowledge of a period is useful for ruling out uniform distribution, since if we have UD(mod N), then the period $\pmod N$ must be divisible by $N$.

D.D.WALL [60] considered the period $k(N)$ of the Fibonacci sequence $\pmod N$ and showed that if $p$ is a prime and $k(p^2) \ne k(p)$ then for $j=1,2,\ldots$ one has $k(p^j) = p^{j-1}k(p)$. He checked also, that for all primes up to $10^4$ one has $k(p^2) \ne k(p)$ and asked whether this holds in general. S.E.MAMANGAKIS [61] proved that this holds for all those primes $p$ for which there exists an element of the Fibonacci sequence divisible by $p$ but not by $p^2$. He gave also a necessary and sufficient condition for validity of $k(p^2) \ne k(p)$ which was later generalized to other sequences by D.W.ROBINSON [66]. P.BUNDSCHUH, J.S.SHIUE [74] proved the first result of Wall for arbitrary second-order recurrences and $p \ne 2$.

2. The first results concerning UD(mod N) for recurrent sequences concerned the Fibonacci sequence and were obtained by L.KUIPERS, J.S. SHIUE [71], [72a] and H.NIEDERREITER [72a]. The problem of characterizing integers $N$ for which a given second-order recurrence is UD(mod N) was first tackled by L.KUIPERS, J.SHIUE [72d] and P.BUNDSCHUH, J.S.SHIUE [73] who obtained sufficient conditions for UD with respect to prime powers, and then solved independently by R.T.BUMBY [75], M.B.NATHANSON [75] and W.A.WEBB, C.T.LONG [75]. (Nathanson considered only the case of primes and Webb and Long dealt with prime powers.) Later H.NIEDERREITER, J.S.SHIUE [77] provided another proof in the case of primes.

3. G.J.RIEGER [77] obtained a sufficient condition for UD(mod m) for recurrences, whose associated polynomial is of the form

$$(x-1)^2(x-m_1)\ldots(x-m_s)$$

where $m_1, m_2, \ldots, m_s$ are distinct integers and his result was extended by L.KUIPERS [79]. A far-reaching generalization was proved by H.NIEDERREITER [80]: let $f(x_1,\ldots,x_k)$ be an integer-valued function defined for integers and satisfying the following condition, if $d|N$ and $x_i \equiv y_i \pmod d$ then $f(x_1,\ldots,x_k) \equiv f(y_1,\ldots,y_k) \pmod d$. Assume further that $u_n$ is a sequence with the property that $u_{n+k} = f(u_n, u_{n+1}, \ldots, u_{n+k-1})$ holds for all $n > 0$. If for every $d|N$, $d \neq 1$ the sequence $u_n \pmod d$ has a period not divisible by $d$, then the sequence $v_n = au_n + n$ is UD(mod N) for every a prime to N. Cf. G.J.RIEGER [79].

## Exercises

1. Give a characterization of those subsets X of the set of all positive integers for which there exists a recurrent sequence $\{u_n\}$ of order two such that $M(u_n) = X$.

2. For a given recurrence sequence $\{u_n\}$ denote by $v(m)$ the smallest period of $u_n \pmod m$. Prove the following results, which go back to M.WARD [31a]:

a). If $m = q_1 \ldots q_t$, where $q_i$ are pairwise co-prime prime powers, then $v(m) = \text{l.c.m.}(v(q_1), \ldots, v(q_t))$.

b). For any prime there exists $k \geq 1$ such that

$$v(p^n) = \begin{cases} v(p) & n = 1, 2, \ldots, k \\ p^{n-k} v(p) & n \geq k. \end{cases}$$

c). The number k occuring in (ii) equals the maximal integer s with the property, that $X^{v(p)} - 1$ lies in the ideal of $Z[X]$ generated by the polynomial associated with $u_n$ and $p^s$.

3. (H.NIEDERREITER, J.S.SHIUE [77]). Prove that if $\{u_n\}$ is a recurrent sequence of second order, then it is UD(mod p) if and only if p divides the discriminant of the associated polynomial, and the sequence $v_n = u_n \pmod p$ does not satisfy a recurrential relation of order smaller than two.

# CHAPTER IV

# ADDITIVE FUNCTIONS

## § 1. The criterion of Delange

1. In this chapter we shall treat sequences $f(1), f(2), \dots$ with $f(n)$ being an *additive function*, i.e. a function satisfying $f(mn) = f(m) + f(n)$ for all pairs $m, n$ of relatively prime integers. Obviously such a function is completely determined by its values at prime powers, thus one would like to have a criterion for UD(mod N) for $f$ in terms of the values $f(p^k)$ ($p$ - prime, $k=1,2,\dots$). Such a criterion was established by H.DELANGE [69] with the aid of a deep result of E.Wirsing, concerning mean values of multiplicative functions. We shall now state a special case of Wirsing's theorem, and then shall deduce the criterion.

PROPOSITION 4.1. (E.WIRSING [67]). *Let* $g(n)$ *be a complex-valued multiplicative function, i.e. a function satisfying* $g(mn) = g(m)g(n)$ *for all pairs* $m, n$ *with* $(m,n) = 1$. *Assume moreover that* $|g(n)| \le 1$ *and that all values* $g(p)$ *for prime* $p$ *lie in a convex polygon contained in the unit circle and containing* 0. *If either the series*

$$\sum_p (1 - \mathrm{Re}\, g(p)) p^{-1} \tag{4.1}$$

*diverges, or, for all* $k \ge 1$ *one has*

$$g(2^k) = -1 ,$$

*then the mean value* $M(g)$ *of* $g$ *defined by*

$$M(g) = \lim_{x \to \infty} \frac{1}{x} \sum_{n \le x} g(n)$$

*exists and equals zero. If none of these conditions holds, then* $M(g)$ *exists and is non-zero.*

For the very long and complicated proof we refer the reader to the original paper (E.WIRSING [67]).

The proof of the criterion of Delange is based on the following consequence of Proposition 4.1:

COROLLARY. *Let* $f(n)$ *be an additive, integer-valued function and let* $N$ *be a given positive integer. Then the function*

$$g(n) = \exp(2\pi i f(n)/N)$$

*has always a mean-value, and this mean-value vanishes if and only if either the series*

$$\sum_{N \nmid f(p)} p^{-1} \tag{4.2}$$

*diverges, or for all* $k \geq 1$ *the number* $2f(2^k)$ *is divisible by* $N$ *and the resulting ratio is odd.*

*Proof*. Since $f$ is additive, the function $g$ is multiplicative and because the values of $g$ are $N$-th roots of unity, the assumptions of the proposition are satisfied. Observe now that $1 - \mathrm{Re}\, g(p)$ exceeds a positive constant, except, when $f(n)/N$ is an integer, thus the series (4.1) divierges if and only if (4.2) diverges. Finally note, that $g(2^k) = -1$ holds for $k \geq 1$ if and only if with suitable integral $M_k$ $(k \geq 1)$ one has

$$2\pi i \frac{f(2^k)}{N} = \pi i + 2\pi i M_k$$

thus $2f(2^k)/N = 1 + 2M_k$ is an odd integer. □

2. To state the necessary and sufficient condition for UD of additive functions we introduce the following two conditions for every integer $d > 1$:

$(A_d)$ The series $\sum_{\substack{p \\ d \nmid f(p)}} p^{-1}$ diverges, and

($B_d$) For all $k \geq 1$ the number $2f(2^k)/d$ is an odd integer.

THEOREM 4.2. (H.DELANGE [69]). *An integer-valued additive function* $f(n)$ *is* UD(mod N) *if and only if for every divisor* $d > 1$ *of* $N$ *either* ($A_d$) *or* ($B_d$) *holds.*

*Proof.* By Proposition (1.1) $f$ will be UD(mod N) if and only if for every $r=1,2,\ldots,N-1$ the function $g_r(n) = \exp(2\pi i r f(n)/N)$ has a zero mean-value. Applying the Corollary to Proposition 4.1 to the function $rf(n)$ we obtain that this will take place if and only if either the series

$$\sum_{N \nmid rf(p)} p^{-1} \tag{4.3}$$

diverges, or, for all $k \geq 1$, the number $2rf(2^k)/N$ is an odd integer. Now we shall transform these conditions. For $r \in [1,N-1]$ write $N = d(r,N)$, $r = r_1(r,N)$ with $(d,r_1) = 1$. Clearly the conditions $N \nmid rf(p)$ and $d \nmid f(p)$ are equivalent, hence the series (4.3) diverges if and only if the condition ($A_d$) holds for $d = N/(r,N)$. Further, if $2rf(2^k)/N = 2r_1 f(2^k)/d$ is an odd integer, then $d$ must be even, $r_1$ odd and so $2f(2^k)/d$ must be an odd integer. Conversely, if $2f(2^k)/N$ is an odd integer. It follows that the condition ($B_d$) is for $d = N/(r,N)$ equivalent with the divisibility of $2rf(2^k)$ by $N$ and the oddness of the resulting ratio.

The Theorem results now from the observation that if $r$ runs over the integers $1,2,\ldots,N-1$, then $d = N/(r,N)$ covers all divisors of $N$ distinct from unity. □

COROLLARY 1. *An additive function* $f$ *is* UD(mod N) *if and only if either* ($A_p$) *holds for every prime divisor* $p$ *of* $N$, *or* $N$ *is even.* ($A_p$) *holds for all odd prime divisors* $p$ *of* $N$, *the numbers* $f(2), f(2^2), \ldots$ *are all odd and in case* $4|N$ *the condition* ($A_4$) *holds as well.*

*Proof.* Assume, that $f$ is UD(mod N) but ($A_p$) fails for a prime divisor $p$ of $N$. Then ($B_p$) must hold, thus $2f(2)/p$ is odd and $p = 2$ follows. Hence ($B_2$) holds and so $f(2^k)$ is odd for all $k \geq 1$. Since $4/2f(2)$ the condition ($B_4$) fails and we see that ($A_4$) must hold.

To get the converse implication observe that if $(A_d)$ holds and $d|D$, then $(A_D)$ holds as well, thus our assumptions imply $(A_d)$ for all divisors $d$ on $N$ except, possibly, $d=2$, but then $(B_2)$ holds and so we may invoke the theorem. □

COROLLARY 2. *If* $f$ *is an additive function, attaining at primes a constant value* $c$, *then* $f$ *will be* UD(mod N) *if and only if* $(N,c)=1$.

*Proof.* If $(N,c)=1$, then $(A_d)$ holds for all $d|N$, $d\neq 1$ hence UD(mod N) follows from the theorem. Assume conversely that $f$ is UD(mod N) but $(N,c)\neq 1$ and let $p$ be any prime divisor of $(N,c)$. Since $(A_p)$ does not hold, the previous corollary implies $p=2$ and $2/f(2)=c$, but $c$ is divisible by $(N,c)$ and hence by 2, so we get a contradiction.

COROLLARY 3. (S.S.PILLAI [40]. In the case $N=2$ H.v.MANGOLDT [98] indicated a proof, details of which were given by E.LANDAU [09], § 167.) *The functions* $\omega(n)$ *and* $\Omega(n)$ *which give the number of distinct prime divisors of* $n$, *repectively the number of prime factors of* $n$, *counted with their multiplicities, are* UD(mod N) *for all* $N$. □

## § 2. Application of Delange's tauberian theorem

1. In this section we shall show that one can obtain the Corollary 3 to the last theorem without the use of Wirsing's theorem, utilizing only classical tools. The same approach works also for a larger class of additive functions, however the reader may convince himself that in this way one cannot obtain a proof of Theorem 4.2 in its full generality.

The main tool here will be the *Ikehara-Delange tauberian theorem*, actually a special case of it, which we now state:

PROPOSITION 4.3. (H.DELANGE [54]). *Let* $a>0$ *and let*

$$f(s) = \sum_{n=1}^{\infty} a_n n^{-s}$$

*be a Dirichlet series with non-negative coefficients, convergent in the open half-plane* $\operatorname{Re} s > a$ *and assume that there exists an integer* $q \geq 1$, *functions* $g_0(s), \dots, g_q(s)$ *regular in the closed half-plane* $\operatorname{Re} s \geq a$ *with* $g_0(a) \neq 0$ *a real number* b *not equal to zero or a negative integer, and complex constants* $a_1, \dots, a_q$ *with real parts smaller than* b. *If for* $\operatorname{Re} s > 1$

$$f(s) = \frac{g_0(s)}{(s-a)^b} + \sum_{j=1}^{q} g_j(s)(s-a)^{-a_j},$$

*then for* x *tending to infinity one has*

$$\sum_{n \leq x} a_n = (C + o(1)) x^a (\log x)^{b-1},$$

*with* $C = g_0(a) a^{-1} \Gamma^{-1}(b)$, $\Gamma$ *denoting the* $\Gamma$*-function.*

For the proof of the general tauberian theorem, of which this is a special case we refer the reader either to the original paper of DELANGE or to chapter III of W.NARKIEWICZ [83a].

2. Now we deduce from Proposition 4.3 (following H.DELANGE [56]) the uniform distribution (mod N), for all N, of the function $\omega(n)$ and $\Omega(n)$. Let f denote either $\omega$ or $\Omega$, and consider for a fixed value of z, $|z| \leq 1$, the Dirichlet series

$$F_z(s) = \sum_{n=1}^{\infty} z^{f(n)} n^{-s}$$

which converges for $\operatorname{Re} s > 1$. Since $z^{f(n)}$ is multiplicative and for prime p $f(p) = 1$, we have

$$F_z(s) = \prod_p (1 + zp^{-s} + \sum_{j=2}^{\infty} z^{f(p^j)} p^{-js}) = g_1(s,z) \prod_p (1 + zp^{-s}),$$

where

$$g_1(s,z) = \prod_p (1 + (\sum_{k=2}^{\infty} z^{f(p^k)} p^{-ks})(1 + zp^{-s})^{-1}).$$

LEMMA 4.4. *The function* $g_1(s,z)$ *is regular for* $\mathrm{Re}\, s > \frac{1}{2}$ *and vanishes at* $s=1$ *only in the case, when* $f=\omega$ *and* $z=-1$.

*Proof.* If $\sigma$ denotes the real part of $s$, then

$$\Big|\Big(\sum_{k=2}^{\infty} z^{f(p^k)} p^{-ks}\Big)(1+zp^{-s})^{-1}\Big| \le \Big(\sum_{k=2}^{\infty} p^{-k\sigma}\Big)(1-p^{-\sigma})^{-1} = (p^{\sigma}-1)^{-2} \le Bp^{-2\sigma}$$

with a certain constant $B$, and since the series

$$\sum_p p^{-2\sigma}$$

converges almost uniformly for $\sigma > \frac{1}{2}$ the first assertion results.

To obtain the second, observe that $g_1(1,z)$ vanishes only if vanishes one of the factors in the product defining it however the p-th factor has its absolute value equal to at least $1-(p-1)^{-2}$ which is positive for $p \neq 2$. Now the factor corresponding to $p=2$ vanishes if and only if

$$\sum_{k=2}^{\infty} \frac{z^{f(2^k)}}{2^k} + \frac{1}{2} z + 1 = 0$$

and this is clearly possible only if $z^{f(2^k)} = -1$ for $k=1,2,\ldots$ and $z=-1$. Since $\Omega(2^2)=2$, the function $f=\Omega$ is excluded, and for $f=\omega$ we have for $z=-1$ indeed $z^{f(2^k)} = -1$ for all $k \ge 1$. □

3. The next lemma deals with the second factor in (4.4).

LEMMA 4.5. *For* $|z| \le 1$ *and* $\mathrm{Re}\, s > 1$ *one has*

$$\prod_p (1+zp^{-s}) = g_2(s,z)(s-1)^{-z}$$

*where* $g_2(s,z)$ *is regular for* $\mathrm{Re}\, s \ge 1$ *and does not vanish at* $s=1$.

*Proof.* For $\mathrm{Re}\, s > 1$ we have

$$\prod_p (1+zp^{-s}) = \exp \sum_p \log(1+zp^{-s}) = \exp\Big(\sum_p (zp^{-s} + \sum_p \sum_{k=2}^{\infty} \frac{(-1)^{k+s}}{k} z^k p^{-ks})\Big) .$$

Since $\sum_p p^{-s} = \log(\frac{1}{s-1}) + g(s)$ with $g(s)$ regular for $\mathrm{Re}\ s \geq 1$ and

$$\left|\sum_{k=2}^{\infty} \frac{(-1)^{k+1}}{k} z^k p^{-ks}\right| \leq \frac{1}{p^{2\sigma} - p^{\sigma}}$$

for $\mathrm{Re}\ s = \sigma > \frac{1}{2}$ we obtain that for $\mathrm{Re}\ s > 1$

$$\prod_p (1 + zp^{-s}) = (s-1)^{-z}\exp\{zg(s)\}g_3(s,z)$$

with $g_3(s,z)$ regular for $\mathrm{Re}\ s > \frac{1}{2}$ and non-vanishing there. □

COROLLARY. *For* $|z| \leq 1$, $z \neq -1$ *and* $\mathrm{Re}\ s > 1$ *one can write*

$$F_z(s) = (s-1)^{-z}H_z(s)$$

*where the function* $H_z(s)$ *is regular for* $\mathrm{Re}\ s \geq 1$ *and* $H_z(1) \neq 0$. *For* $z = -1$ *the function* $F_z(s)$ *is regular for* $\mathrm{Re}\ s \geq 1$. □

4. After these preliminaries we can now return to our main task. Given a positive integer $N$, put $z_r = \exp(2\pi ir/N)$ for $r=0,1,\ldots,N-1$, fix $j$, and observe that for $\mathrm{Re}\ s > 1$

$$\sum_{\substack{n \\ f(n)\equiv j \pmod N}} n^{-s} = \frac{1}{N}\sum_{r=0}^{N-1} F_{z_r}(s)\exp\{-2\pi ijr/N\} .$$

Indeed, this follows by substituting here the series for $F_{z_r}(s)$ and interchanging the summations. Using the Corollary to the last Lemma we obtain

$$\sum_{\substack{n \\ f(n)\equiv j \pmod N}} n^{-s} = \frac{1}{N} \cdot \frac{H_1(s)}{s-1} + \sum_{r=1}^{N-1} \frac{H_{z_r}(s)}{(s-1)^{z_r}} \exp\{-2\pi ijr/N\}$$

Since $\mathrm{Re}\ z_r < 1$ for $r=1,2,\ldots,N-1$ the assertion about UD(mod N) for the function $f$ results from Proposition 4.3 since $H_1(1) \neq 0$.

In the next chapter we shall apply the same Proposition 4.3 to problems concerning weak uniform distribution.

## § 3. The sets M(f) for additive functions

Let us now consider the question of characterizing the sets $M(f)$ for additive functions $f$. The answer to it is given by the following

THEOREM 4.6. *If* $X$ *is a non-empty set of positive integers, then there exists an additive function* $f$ *satisfying* $M(f)=X$ *if and only if for a suitable set of primes* $P$ *(which may be empty) we have either*

$$X = \{\prod_{p\in P} p^{\alpha_p} : \alpha_p \ge 0\} = A(P)$$

or

$$X = \{2^{\alpha} \prod_{p\in P} p^{\alpha_p} : \alpha_p \ge 0,\ 0 \le \alpha \le 1\} = B(P) .$$

*Proof*. Observe first that from the Corollary 1 to Theorem 4.2 it follows that if $f$ is UD(mod M) and UD(mod N) and not both numbers $M,N$ are even, then $f$ is also UD(mod MN). Moreover note, that the same corollary implies that if $f$ is UD(mod 4), then it is also UD(mod $2^m$) for $m=1,2,\dots$ . The necessity of the stated condition results now immediately.

To prove its sufficiency we shall now construct explicitly additive functions $f$, for which $M(f)$ will have the form stated in the Theorem. Let thus $P$ be a given set of primes. If it is empty, then $M(f)=A(P)$ holds for $f(n)=0$ and $M(f)=B(P)$ holds for the function defined by

$$f(n) = \begin{cases} 0 & \text{if } n=0 \text{ or } 2\nmid n \\ 1 & \text{if } 2|n . \end{cases}$$

If $P$ consists of all primes, then $A(P)=B(P)$ and $f(n)=n$ realizes our needs. So assume that neither $P$ nor its complement $P'$ in the set of all primes is void. Let $q_1 < q_2 < \dots$ be the sequence of all primes in $P'$ and let $3=p_1 < p_2 < \dots$ be the sequence of all odd primes. If $X=A(P)$ then we define a completely additive function $f$ (i.e. satisfying $f(ab)=f(a)+f(b)$ for all $a,b$) by putting $f(2)=2$ and

$$f(p_j) = q_1 \dots q_j \qquad (j=1,2,\dots) .$$

In case when $P'$ is finite and consists of, say, $T$ primes, we define formally $q_{T+1} = q_{T+2} = \dots = 1$, so that our definition of $f$ makes sense. For this function the condition $(A_p)$ holds for every prime $p$ in $P$ and for none in $P'$. Since $(B_2)$ evidently fails the Corollary 1 to Theorem 4.2 gives $M(f) = A(P)$.

If $X = B(P)$ (in which case we may assume $2 \in P'$) we define $f$ by putting

$$f(2^n) = 1 \qquad (n=1,2,\dots),$$

$$f(p_j^n) = 2q_1q_2\dots q_j \qquad (n=1,2,\dots;\ j=1,2,\dots)$$

(with the same convention as before in the case of finite $P'$). The condition $(A_p)$ holds for all primes in $P$ and for none in $P'$, however $(B_2)$ holds and since $(A_4)$ fails (due to $q_1 = 2$ all values $f(p)$ for odd $p$ are divisible by 4) we may again use the Corollary 1 to Theorem 4.2 to get $M(f) = B(P)$. □

## § 4. Notes and comments

1. The first results in the direction of Theorem 4.2 occur already in H.DELANGE [61], who proved a weaker version of Wirsing's Theorem. He could assert the vanishing of $M(f)$ for $f$ multiplicative and absolutely bounded by unity in the case when

$$\sum_{p\le x} f(p) = (c + o(1)) \frac{x}{\log x}$$

and in the case $c = 1$ the series

$$\sum_p \frac{1 - \operatorname{Re} f(p)}{p}$$

diverges.

2. The main result of this chapter, Theorem 4.2 was proved by H. DELANGE [69] who in the same paper considered also joint distribution

of the pair $\langle f(n) \bmod N_1, g(n) \bmod N_2\rangle$, where $f,g$ are both additive and showed that if $(N_1,N_2)=1$ and $f$ is UD(mod $N_1$), $g$ is UD(mod $N_2$), then for arbitrary $a,b$ one has

$$\lim_{x\to\infty} \frac{1}{x} \#\{n \le x:\ f(n) \equiv a \pmod{N_1},\ g(n) \equiv b \pmod{N_2}\} = \frac{1}{N_1 N_2}$$

i.e. the pair $f,g$ is UD with respect to $N_1,N_2$.

In a later paper (H.DELANGE [74]) he considered the case $(N_1,N_2) = d \neq 1$ and showed that a pair $f,g$ of additive functions will be UD with respect to $N_1,N_2$ if and only if $f$ is UD(mod $N_1$), $g$ is UD(mod $N_2$) and moreover for every pair $t_1,t_2$ of coprime integers satisfying $0 < t_1 < d$, $0 < t_2 < d$ the function $t_1 f + t_2 g$ is UD(mod d). He noted also, that this condition is necessary for arbitrary, not necessarily additive, functions $f,g$. A corresponding result is also true for systems of more than two additive functions.

3. In H.DELANGE [72] an analogue of Theorem 4.2 was obtained for sequences of values of an additive function restricted to a set $A$, which is assumed to be an intersection of an arithmetical progression with a set $M$ with multiplicative characteristic function and with $\sum_{p \notin M} 1/p < \infty$. In particular $M$ may contain all integers, thus one obtains criterias for UD(mod N) for functions of the form $f(an+b)$ with additive $f$ and given $a,b$.

Recently it was shown by H.DABOUSSI, H.DELANGE [82] (th.7 (iii)) that if $x$ is an irrational real number and $f$ an additive function which is UD(mod N), then the sequence $f([nx])$ is also UD(mod N).

## Exercises

1. Prove that if $N$ is odd, $f,g$ are additive and $f+g$ is UD(mod N) then either $f$ or $g$ must be UD(mod N).

2. Characterize the sets $P$ of primes for which the function $\omega_P(n) = \sum_{\substack{p \mid n \\ p \in P}} 1$ is UD(mod N) for all $N$.

3. (H.DELANGE [69]). Prove, that if $f,g$ are additive, $f$ is UD(mod $N_1$), $g$ is UD(mod $N_2$) and $(N_1,N_2)=1$, then the pair $f,g$ is UD with respect to $N_1$ and $N_2$.

4. Show that the assertion of the previous exercise may be false if $(N_1, N_2) \neq 1$.

5. Obtain a criterion for uniform distribution for a system of additive functions.

6. Let $A, B$ be two sets of primes, and let $\omega_A$, $\omega_B$ denote the number of distinct prime factors of $n$ lying in $A$ resp. $B$. Determine, when the pair $\omega_A(n)$, $\omega_B(n)$ is uniformly distributed with respect to given $M, M$.

# CHAPTER V

# MULTIPLICATIVE FUNCTIONS

## § 1. Dirichlet - WUD

1. In this chapter we shall study multiplicative functions. Here the appropriate notion is that of weak uniform distribution and we shall be aiming at necessary and sufficient conditions for WUD(mod N) for a possibly large class of multiplicative functions. We shall first consider a weaker notion of uniform distribution than WUD, obtain a necessary and sufficient condition for it to hold for a given multiplicative function and then utilize tauberian theorem of Delange (our Proposition 4.3) to deduce a criterion for WUD(mod N) for a reasonably large class of multiplicative functions containing all those which are *"polynomial-like"*, i.e. satisfy

$$f(p^j) = V_j(p)$$

with certain polynomials $V_j(x)$ and all primes $p$ for all or at least for sufficiently many positive indices $j$.

To introduce the new notion, which we shall call *Dirichlet-WUD(mod N)*, or shortly *D-WUD(mod N)*, consider a multiplicative, integer-valued function $f$ and a given integer $N \geq 3$. We assume that there exists a positive integer $r$ with the property that the series

$$\sum_{\substack{p \\ (f(p^r),N)=1}} 1/p$$

diverges. Let $m = m(f,N)$ be the smallest integer with this property.

We shall say that the function $f$ is *Dirichlet-WUD(mod N)* if and only if for every $j$ prime to $N$ one has

$$\lim_{s\to 1/m+0} \Big( \sum_{\substack{n \\ f(n)\equiv j (\mathrm{mod}\ N)}} n^{-s} \Big) : \Big( \sum_{\substack{n \\ (f(n),N)=1}} n^{-s} \Big) = 1/\varphi(N) . \tag{5.1}$$

To guarantee that this definition is correct we have to be sure that the abscissas of absolute convergence of both Dirichlet series occuring here are at most equal to $1/m$ and it suffices to do this for the second of them.

Denoting by $\varepsilon(n)$ the (multiplicative) characteristic function of the set of all those $n$'s for which $(f(n),N)=1$ we can write

$$\sum_{\substack{n \\ (f(n),N)=1}} n^{-s} = \sum_{n=1}^{\infty} \varepsilon(n) n^{-s} = \sum_{n\in\bar{A}} \varepsilon(n) n^{-s} \sum_{n\in\bar{B}} \varepsilon(n) n^{-s}$$

at every $s$, where the two series on the right converge absolutely, with $A$ being the set of all primes $p$ for which with a certain $1 \le j \le m-1$ one has $(f(p^j),N)=1$, $B$ being the set of remaining primes and $\bar{A}$, $\bar{B}$ denoting the sets of all integers composed of primes lying in $A$ resp. $B$. Since $\sum_{p\in A} 1/p$ converges, the first factor is absolutely convergent for $\mathrm{Re}\ s > 0$ and in view of

$$\sum_{n\in\bar{B}} \varepsilon(n) n^{-s} = \prod_{p\in B} (1 + \varepsilon(p^m) p^{-ms} + \ldots)$$

we conclude that the second factor converges absolutely for $\mathrm{Re}\ s > 1/m$.

2. To formulate the criterion for D-WUD(mod N) denote by $A_j$, for $j$ prime to $N$, the set of all primes $p$ satisfying $f(p^m) \equiv j (\mathrm{mod}\ N)$ and denote by $\Lambda = \Lambda(f,N)$ the subgroup of $G(N)$ (the multiplicative group of residue classes (mod N),,prime to N) generated by those residues $j(\mathrm{mod}\ N)$ for which the series

$$\sum_{p\in A_j} p^{-1}$$

diverges.

THEOREM 5.1. *A multiplicative integer-valued function* f, *for which* m(f,N) *is well-defined is* D-WUD(mod N) *if and only if for every character* X(mod N), *trivial on* $\Lambda$ *there exists a prime* p *such that with* $m = m(f,N)$

$$\sum_{j=0}^{\infty} X(f(p^j))p^{-j/m} = 0 . \qquad (5.2)$$

(Note, that necessarily $p \le 2^m$, since

$$\left|\sum_{j=0}^{\infty} X(f(p^j))p^{-j/m}\right| \ge 1 - \sum_{j=1}^{\infty} p^{-j/m} = 1 - \frac{1}{p^{1/m} - 1}$$

which is positive for $p > 2^m$.)

*Proof*. We shall transform the formula (5.1) and to do this we have to look more carefully at the series

$$\sum_{\substack{n \\ f(n) \equiv j \pmod s}} n^{-s}$$

occuring in it. Note first that it equals to

$$\frac{1}{\varphi(N)} \sum_{X} \overline{X(j)}\, F(s,X) , \qquad (5.3)$$

where the summation is taken over all characters X(mod N) and

$$F(s,X) = \sum_{n=1}^{\infty} X(f(n))n^{-s}$$

is a series convergent absolutely for $\mathrm{Re}\, s > 1/m$. Since X(f(n)) is multiplicative we can utilize Euler's products to obtain:

LEMMA 5.2. *For every character* X(mod N) *one has in the half-plane* $\mathrm{Re}\, s > 1/m$

$$F(s,X) = (s - 1/m)^{\alpha(X)} g(s,X) \exp\{\sum_{p} X(f(p^m))p^{-ms}\} ,$$

*where* $\alpha(X)$ *is a non-negative integer, which is positive if and only if for some prime* $p$ *the equality (5.2) is true, and the function* $g(s,X)$ *is regular in the closed half-plane* $\mathrm{Re}\, s \geq 1/m$ *and does not vanish at* $s = 1/m$.

*Proof.* If we denote the series $\sum_{k=0}^{\infty} X(f(p^k))p^{-ks}$ by $T_p(s)$, then for $\mathrm{Re}\, s > 1/m$ we can write

$$F(s,X) = \prod_p T_p(s) = \prod_{p \leq 2^m} T_p(s) \prod_{p > 2^m} T_p(s)$$

The first factor is regular for $\mathrm{Re}\, s > 0$ and will vanish at $s = 1/m$ if and only if for a certain prime $p$ we have (5.2). We can thus write it in the form $(s - 1/m)^{\alpha(X)} g_1(s,X)$ with $g_1(s,X)$ regular for $\mathrm{Re}\, s \geq 1/m$ and non-vanishing at $s = 1/m$ and $\alpha(X)$ has the properties listed in the lemma. Since for $p > 2^m$ and $\mathrm{Re}\, s \geq 1/m$ we have

$$|T_p(s)| \geq 1 - \sum_{k=1}^{\infty} p^{-k/m} = 1 - (p^{1/m} - 1) > 0 \tag{5.4}$$

hence for these $p$'s $T_p(s)$ does not vanish for $\mathrm{Re}\, s \geq 1/m$.

From the product $\prod_{p>2^m} T_p(s)$ we separate now the part corresponding to those primes $p$, for which there is an index $1 \leq j \leq m-1$ such that $(f(p^j), N) = 1$. By our assumption the series of inverses of those $p$'s converges and hence the separated part, which we denote by $g_2(s,X)$ is regular for $\mathrm{Re}\, s > 0$ and does not vanish at $s = 1/m$ due to (5.4). For the remaining primes $p > 2^m$ (whose set we denote by $A$) we have

$$T_p(s) = 1 + \sum_{k=m}^{\infty} X(f(p^k))p^{-ks} \neq 0$$

by (5.4), hence for $\mathrm{Re}\, s > 1/m$ we can write

$$\prod_{p\in A} T_p(s) = \exp \sum_{p\in A} \log T_p(s) = \exp\{ \sum_{p\in A} X(f(p^m))p^{-ms} + g_3(s,X)\}$$

with a certain function $g_3(s,X)$ regular for $\mathrm{Re}\, s \geq 1/m$.

Putting everything together we arrive at

$$F(s,X) = (s-1/m)^{\alpha(X)} g_1(s,X) g_2(s,X) \exp\{ \sum_{p\in A} X(f(p^m)) p^{-ms} + g_3(s,X)\} ,$$

thus putting

$$g_4(s,X) = \exp\{-\sum_{p\notin A} X(f(p^m)) p^{-ms}\}$$

and $g(s,X) = g_1(s,X) g_2(s,X) \exp(g_3(s,X) + g_4(s,X))$ we obtain our assertion. □

Since for the principal character $X_0 \pmod N$ one has

$$F(s,X_0) = \sum_{\substack{n \\ (f(n),N)=1}} n^{-s}$$

we obtain that $f$ will be D-WUD(mod N) if and only if for every $j$ prime to $N$ one has

$$\frac{1}{\varphi(N)} = \lim_{s\to\frac{1}{m}+0} \frac{\frac{1}{\varphi(N)} \sum_X \overline{X(j)}\, F(s,X)}{F(s,X_0)} = \frac{1}{\varphi(N)} \lim_{s\to\frac{1}{m}+0} \Big(1 + \sum_{X\neq X_0} \overline{X}(j) \frac{F(s,X)}{F(s,X_0)}\Big) =$$

$$= \frac{1}{\varphi(n)} + \frac{1}{\varphi(N)} \sum_{X\neq X_0} \overline{X}(j) \lim_{s\to\frac{1}{m}+0} \frac{F(s,X)}{F(s,X_0)}$$

and using the lemma and the obvious equality $\alpha(X) = 0$ we see that this is equivalent to

$$\sum_{X\neq X_0} \overline{X(j)} g(1/m,X) \lim_{s\to\frac{1}{m}+0} \left(s-\frac{1}{m}\right)^{(X)} \exp\left\{ \sum_{\substack{k \\ (k,N)=1}} (X(k)-1) \sum_{p\in A_k} p^{-ms} \right\} = 0 \qquad (5.5)$$

holding for all $j$ prime to $N$. However, since the matrix

$$(\overline{X(j)})_{\substack{X\neq X_0 \\ (j,N)=1}}$$

is of rank $\varphi(N)-1$ it follows that (5.5) holds if and only if for all

$X \neq X_0$ one has

$$\lim_{s\to\frac{1}{m}+0} \left\{ \sum_{\substack{k\\(k,N)=1}} (\mathrm{Re}\ X(k)-1) \sum_{p\in A_k} p^{-sm} + \alpha(X)\log(s-\tfrac{1}{m}) \right\} = -\infty \qquad (5.6)$$

Assume now that for every non-principal character which is trivial on $\Lambda$ there exists a prime $p$ such that (5.2) holds. If $X$ is such, then $\alpha(X) \geq 1$ and since $\mathrm{Re}\ X(k) \leq 1$ we obtain (5.6). If however $X$ is not trivial on, then we may select $r$ in $\Lambda$ with $X(r) \neq 1$. For such $r$, $\mathrm{Re}\ X(r) < 1$ and since for remaining $k$'s $\mathrm{Re}\ X(k) \leq 1$ we obtain for $s > 1/m$

$$\sum_{\substack{k\\(k,N)=1}} (\mathrm{Re}\ X(k)-1) \sum_{p\in A_k} p^{-sm} + \alpha(X)\log(s-\tfrac{1}{m}) \leq (\mathrm{Re}\ X(r)-1) \sum_{p\in A_k} p^{-sm} \to -\infty .$$

Thus $f$ is D-WUD(mod N).

To obtain the converse implication assume that $f$ is D-WUD(mod N) but there exists a non-principal character $X$, trivial on $\Lambda$ for which $\alpha(X) = 0$. Then for $k$ in $\Lambda$ we have $X(k) = 1$ and for the remaining $k$'s the function

$$\sum_{p\in A_k} p^{-sm}$$

is regular at $s = 1/m$, thus (5.6) cannot hold, contradicting D-WUD(mod N). □

COROLLARY. *If* $\Lambda(f,N) = G(N)$, *then* $f$ *is* D-WUD(mod N). *Moreover, if* $q$ *is an odd prime and* $\Lambda(f,q^2) = G(q^2)$ *then for all* $k \geq 1$ *we have* $\Lambda(f,q\ ) = G(q\ )$ *and thus* $f$ *is* D-WUD(mod $q^k$). *In the case* $q = 2$ *the same assertion is valid, however under a slightly stronger assumption, namely* $\Lambda(f,2^3) = G(2^3)$.

*Proof.* The first part follows immediately from the theorem. To obtain the second note that in view of $\Lambda(f,q^2) = G(q^2)$ we may find residue classes $j_1,\dots,j_t$ (mod $q^2$) such that all series

$$\sum_{\substack{p\\ f(p^m)\equiv j_i \pmod{q^2}}} 1/p \qquad (i=1,2,\dots,t)$$

(with $m = m(f,q^2) = m(f,q^k)$ $(k \geq 1)$) diverge and the product $j = j_1 \dots j_t$ is a primitive root $\pmod{q^2}$. For given $k \geq 2$ fix $1 \leq i \leq t$ and consider those residue classes $r \pmod{q^k}$ which satisfy $r \equiv j_i \pmod{q^2}$. There must exist at least one of them, say $r_i$, with the property that the series

$$\sum_{\substack{p \\ f(p^m) \equiv r_i \pmod{q^k}}} 1/p$$

diverges, so that $r \in \Lambda(f,q^k)$. Thus $r = r_1 \dots r_k$ lies also in $\Lambda(f,q^k)$ but $r \equiv j_1 \dots j_k \equiv j \pmod{q^2}$, hence $r$ is a primitive root $\pmod{q^2}$ and thus also $\pmod{q^k}$, implying $\Lambda(f,q^k) = G(q^k)$. The argument in case $q = 2$ is analogous, one has only to remember that $G(2^k)$ has for $k \geq 3$ two generators: $5 \pmod{2^k}$ and $-1 \pmod{2^k}$, and if $g_1 \equiv 5 \pmod 8$, $g_2 \equiv -1 \pmod 8$, then the pair $g_1, g_2$ generates $G(2^k)$. □

3. It is now time to show how D-WUD(mod N) is connected with WUD(mod N). We prove:

PROPOSITION 5.3. *If* f *is a multiplicative, integer-valued function which is* WUD(mod N) *then it is also* D-WUD(mod N).

*Proof*. The assertion results immediately from the following Lemma, which is due essentially to R.DEDEKIND [94], who treated the case $a_n = 1$.

LEMMA 5.4. *Let* $f(s) = \sum_{n=1}^{\infty} a_n n^{-s}$, $g(s) = \sum_{n=1}^{\infty} b_n n^{-s}$ *be two Dirichlet series with* $a_n$ *non-negative and* $b_n$ *arbitrary complex numbers, which have* a *resp.* b *for their abscissas of convergence, with* $0 < a \leq b$ *and assume further, that* $|f(s)|$ *tends to infinity, when* s *tends to* b *over real values* $> b$. *Put further* $A(x) = \sum_{n \leq x} a_n$, $B(x) = \sum_{n \leq x} b_n$. *If*

$$\lim_{x \to \infty} B(x)/A(x) = 1$$

*then* $a = b$ *and*

$$\lim_{s \to a+0} f(s)/g(s) = 1 .$$

*Proof*. The first assertion results from the following majoration valid for $s > a$ and $x \ge 2$

$$\left|\sum_{n\le x} b_n n^{-s}\right| = \left|\sum_{n\le x} B(n)\left(\frac{1}{n^s} - \frac{1}{(n+1)^s}\right) + B(x)[x]^{-s}\right| \le$$

$$\le C\left\{\sum_{n\le x} A(n)\left(\frac{1}{n^s} - \frac{1}{(n+1)^s}\right) + A(x)[x]^{-s}\right\} = C\sum_{n\le x} a_n n^{-s}$$

which holds with a certain constant $C$, and once we know that the abscissas the convergence coincide we use the equalities

$$f(s) = \sum_{n=1}^{\infty} A(n)\left(\frac{1}{n^s} - \frac{1}{(n+1)^s}\right) = \sum_{n=1}^{\infty} A(n)\, s \int_{\log n}^{\log(n+1)} e^{-st}dt =$$

$$= \sum_{n=1}^{\infty} s \int_{\log n}^{\log(n+1)} e^{-st} A(e^t)dt = s\int_0^{\infty} e^{-st} A(e^t)dt \tag{5.7}$$

and

$$g(s) = s\int_0^{\infty} e^{-st} B(e^t)dt$$

to get, with $R(x) = B(x) - A(x) = o(A(x))$,

$$g(s) = f(s) + s\int_0^{\infty} e^{-st} R(e^t)dt$$

and if $|R(x)/A(x)| < \varepsilon$ holds for $x \ge x_0$, then with suitable $C_1$

$$\left|\int_0^{\infty} e^{-st}R(e^t)dt\right| \le \varepsilon s \int_{\log x_0}^{\infty} e^{-st}A(e^t)dt + sC_1\int_0^{\log x_0} e^{-st}dt \le \varepsilon s f(s) + O(1)$$

implying $g(s) = f(s) + o(f(s))$.

COROLLARY. *If* $N \ge 3$ *and* $f$ *is an integer-valued multiplicative function for which* $m(f,N)$ *is defined and which is* WUD(mod N), *then it satisfies the conditions given in Theorem 4.1.* □

4. It seems that no example is known of a multiplicative function which is D-WUD(mod N) for a certain N, without being WUD(mod N). Hence the following

PROBLEM III. *Are the notions* WUD(mod N) *and* D-WUD(mod N) *equivalent ?*

One can however find large classes of functions for which these notions coincide. The first class. which we shall denote by $F_N$, consists of all those multiplicative functions for which the series

$$\sum_{\substack{p \\ (f(p),N)\neq 1}} 1/p$$

converges. For this class our two notions coincide for one modulus namely N, and we shall deduce this from Wirsing's proposition 4.1.

PROPOSITION 5.5. *If* $N \geq 3$ *is a given integer and* $f \in F_N$, *then* f *is* WUD(mod N) *if and only if it is* D-WUD(mod N).

*Proof*. Observe first, that if f is any multiplicative function, not necessarily lying in $F_N$, and $(j,N) = 1$ then

$$\#\{n \leq x:\ f(n) \equiv j \pmod N\} = \frac{1}{\varphi(N)} \sum_{X \pmod N} \overline{X(j)}\, X(f(n)) ,$$

thus

$$d_j = \lim_{x\to\infty} x^{-1} \#\{n \leq x:\ f(n) \equiv j \pmod N\} = \frac{1}{\varphi(N)} \sum_{X \pmod N} \overline{X}(j) M(X(f))$$

where $M(g)$ denotes, as before, the mean value of the function g. The existence of the mean values occuring in this formula is a consequence of Proposition 4.1, since the values of $X(f(n))$ are either zero or roots of unity of bounded degrees. Note now that for $f \in F_N$ the mean value $M(X_0(f))$, with $X_0$ denoting the principal character (mod N), is non-zero by the same Proposition, since the condition (4.1) is violated and $X_0(f(2^k)) \neq 1$. It follows that $d = \sum_{(j,N)=1} d_j$ is positive.

Applying Lemma 5.3 to the series

$$f_j(s) = \sum_{\substack{n \\ f(n)\equiv j \pmod N}} n^{-s}$$

and

$$g(s) = \sum_{\substack{n \\ (f(n),N)=1}} n^{-s}$$

and observing that their common abscissa of convergence equals unity, we obtain that the limit in (5.1) equals

$$\lim_{s\to 1+0} f_j(s)/g(s) = \lim_{x\to\infty}(\#\{n\le x:\ f(x)\equiv j \pmod N\}:\ \#\{n\le x:\ (f(n),N))=1\} = d_j/d\ ,$$

hence D-WUD(mod N) and WUD(mod N) coincide in this case. □

Note however that this Proposition is not particularly useful for deciding whether a given function $f$ from $F_N$ is WUD(mod N) or not, since one can do this applying directly Proposition 4.1 without the use of Theorem 5.1 (see Exercise 3).

## § 2. Decent functions

1. The second class of functions which we shall now consider is much larger and contains most of the multiplicative functions occuring in the nature, like Euler's $\varphi$-function, the sum of divisors, the number of divisors, Ramanujan's $\tau$-function etc.

This class consist of all multiplicative integer-valued functions $f$ which have the following property:

*For every integer* $N \ge 3$ *there exists an integer* $T$ *with the property that for* $r=1,2,\dots,T$ *and for all integers* $j$, *prime to* $N$ *one can write for* $\mathrm{Re}\ s > 1$

$$\sum_{\substack{p \\ f(p^r)\equiv j \pmod N}} p^{-s} = a(r,j)\log \frac{1}{s-1} + g_{r,j}(s)$$

*with* $a(r,j) \geq 0$ *and* $g_{r,j}(s)$ *regular in the closed half-plane* $\mathrm{Re}\, s \geq 1$. *Moreover at least one number* $a(r,j)$ $(r \leq T,\ j \geq 1)$ *should be positive.*

Such functions will be called *decent*, and the maximal possible value of $T$ (possibly infinite) will be called the *order of decency of* $f$ *at* $N$. In dealing with WUD for a particular value of $N$ it will be sufficient to consider these conditions only for this one number, however we shall not have an opportunity to apply this observation.

A broad subclass of decent functions is formed by the *polynomial-like* functions, which we define as those integer-valued multiplicative functions $f$ for which there exists a polynomial $P$ over $Z$ such that for all primes $p$ one has $f(p) = P(p)$. More precisely, we shall speak about a polynomial-like function of order $k$, if there exist polynomials $P_1(x),\dots,P_k(x)$ over $Z$ such that for all primes $p$ one has

$$f(p^j) = P_j(p) \qquad (j=1,2,\dots,k) . \tag{5.8}$$

If this condition is satisfied for all $k$, as it happens for the Euler's $\varphi$-function (with $P_j(x) = x^{j-1}(x-1)$) or for the divisor function (with $P_j(x) = 1+j$), then we say that $f$ is of *infinite order*. The *exact order of* $f$ (denoted by $E(f)$) is defined as the largest integer (or $\infty$) for which $f$ is polynomial-like of order $k$.

PROPOSITION 5.6. *Every polynomial-like function* $f$ *is decent and the order of its decency at* $N$ *is* $\geq E(f)$.

*Proof.* Let $r \leq E(f)$, and let $N$ and $j$, with $(j,N) = 1$ be given. Let $x_1,\dots,x_k$ be all solutions of the congruence $P_r(x) \equiv j \pmod N$ which are prime to $N$, if there exist any. Then

$$\sum_{\substack{p \\ f(p^r)\equiv j \pmod N}} p^{-s} = \sum_{\substack{p \\ P_r(p)\equiv j \pmod N}} p^{-s} = \sum_{j=1}^{k} \sum_{\substack{p \\ p\equiv x_j \pmod N}} p^{-s} = \frac{k}{\varphi(N)} \log \frac{1}{s-1} + g(s)$$

with $g(s)$ regular for $\operatorname{Re} s \geq 1$ by the Dirichlet's Prime Number Theorem. We may hence put $a(r,j) = k/\varphi(N)$. If there are no solutions of that congruence, then the series

$$\sum_{\substack{p \\ P_r(p) \equiv j \pmod N}} p^{-s}$$

reduces to a finite sum and so we may take 0 for $a(r,j)$. □

2. THEOREM 5.7. *If* $f$ *is a decent multiplicative function,* $N \geq 3$ *a given integer and there exists an index* $r$ *not exceeding the order of decency of* $f$ *at* $N$ *such that not all numbers* $a(r,j)$ $((j,N)=1)$ *vanish and* $f$ *is* D-WUD(mod N) *then it is also* WUD(mod N).

*Proof.* We may assume that the index $r$ is chosen as small as possible, thus for $i=1,2,\ldots,r-1$ and all $j$ prime to $N$ the series

$$\sum_{\substack{p \\ f(p^i) \equiv j \pmod N}} p^{-s}$$

represents a function regular for $\operatorname{Re} s \geq 1$, and hence the same applies to

$$\sum_{\substack{p \\ (f(p^i),N)=1}} p^{-s}$$

Note also that the series

$$\sum_{\substack{p \\ (f(p^r),N)=1}} 1/p$$

certainly diverges, since otherwise the series

$$\sum_{\substack{p \\ (f(p^r),N)=1}} p^{-s}$$

would represent a function regular at $s=1$, contrary to the choice of $r$. This shows that for $f$ the number $m(f,N)$ coincides with $r$. Observe also that for any character $X \pmod N$ and $\operatorname{Re} s > 1/r$

$$\sum_p X(f(p^r))p^{-rs} = \sum_{(j,N)=1} X(j) \sum_{\substack{p \\ f(p^r)\equiv j \pmod N}} p^{-rs} =$$

$$= \sum_{(j,N)=1} X(j)\left(a(r,j)\log \frac{1}{rs-1} + g_{r,j}(rs)\right) =$$

$$= \left\{\sum_{(j,N)=1} X(j)a(r,j)\right\}\log \frac{1}{s-1/r} + h_j(s)$$

holds with $h_j(s)$ regular for $\mathrm{Re}\, s \geq 1/r$ and thus by Lemma 5.2

$$F_j(s) = \sum_{\substack{n \\ f(n)\equiv j \pmod N}} n^{-s} = \frac{1}{\varphi(N)}\left\{\sum_{X\neq X_0} \overline{X(j)}\, \frac{h_j(s,X)}{(s-1/r)^{\beta(X)}} + \frac{h(s,X_0)}{(s-1/r)^{\beta(X_0)}}\right\}$$

with $\beta(X) = \sum_{(j,N)=1} X(j)a(r,j)$ and $h_j(s,X)$ regular for $\mathrm{Re}\, s \geq 1/r$,

Also by Lemma 5.2 we get

$$F(s) = \sum_{\substack{n \\ (f(n),N)=1}} n^{-s} = \frac{h(s,X_0)}{(s-1/r)^{\beta(X_0)}}$$

Since by assumption

$$\lim_{s\to \frac{1}{r}+0} F_j(s)/F(s) = 1$$

and also $\mathrm{Re}\, \beta(X) \leq \beta(X_0)$ with equality possible only in case $\beta(X) = \beta(X_0)$ it follows that we can write $F_j(s)$ in the form

$$F_j(s) = \frac{1}{\varphi(N)} \frac{g_0(s)}{(s-1/r)^{\beta(X_0)}} + \sum_{j=1}^{q} g_j(s)\left(s-\frac{1}{r}\right)^{-\alpha_j}$$

with $g_0(s),\dots,g_q(s)$ regular for $\mathrm{Re}\, s \geq 1/r$, $\mathrm{Re}\, \alpha_j < \beta(X_0)$ and $g_0(1/r) = h(1/r,X_0)$

Applying finally Proposition 4.3 to $F(s)$ and $F_j(s)$ we arrive at

$$\#\{n\leq x:\ f(n)\equiv j \pmod N\} = \left(r\,\frac{g_0(1/r)}{\varphi(N)\Gamma(\beta(X_0))} + o(1)\right) x \log^{\beta(X_0)-1} x$$

and

$$\#\{n\le x:\ (f(n),N)=1\} = \left(\frac{rg_0(1/r)}{\varphi(N)\Gamma(\beta(X_0))}+o(1)\right) x\log^{\beta(X_0)-1} x$$

which due to the arbitrariness of $j$ proves that $f$ is WUD(mod N). □

We may thus use Theorem 5.1 for checking WUD(mod N) for decent functions.

3. In the case of polynomial-like functions it is useful to have a criterion for WUD(mod N) expressed in terms of the polynomials $P_1, P_2, \dots$ occuring in (5.8). Such a criterion is given by the following

COROLLARY. *Let* $f$ *be multiplicative and polynomial-like of exact order* $E$. *For* $j=1,2,\dots,E$ *denote by* $R_j = R_j(f,N)$ *the subset of* $G(N)$, *the multiplicative group of residue classes* (mod N) *prime to* $N$, *consisting of those residue classes* $r\in G(N)$ *for which the congruence* $P_j(x)\equiv r \pmod N$ *has a solution in* $G(N)$. *Let* $\Lambda_j = \Lambda_j(f,N)$ *be the subgroup of* $G(N)$ *generated by* $R_j$ *and assume further that not all sets* $R_1, R_2, \dots, R_E$ *are empty (in case* $E=\infty$ *we assume that not all sets* $R_i$ *are empty). Let* $m=m(f,N)$ *be the smallest index with* $R_m$ *non-empty. Then the function* $f$ *will be* WUD(mod N) *if and only if for every non-principal character* $X \pmod N$ *trivial on* $\Lambda_m$ *there exists a prime* $p$ *such that*

$$\sum_{j=0}^{\infty} X(f(p^j))p^{-j/m} = 0 .$$

*In particular, if* $\Lambda_m(f,N) = G(N)$ *then* $f$ *is* WUD(mod N) *and if is an odd prime such that* $\Lambda_m(f,p^2) = G(p^2)$ *(with* $m=m(f,p^2)$*) then* $\Lambda_m(f,p^k) = G(p^k)$ *and* $f$ *is* WUD(mod p ) *for* $k\ge 1$. *A similar result holds for* $p=2$, *however one has to assume* $\Lambda_m(f,2^3) = G(2^3)$

*Proof.* Note that the group $\Lambda$ occuring in Theorem 5.1 coincides with $\Lambda_m$. Indeed, the series

$$\sum_{\substack{p \\ f(p^m)\equiv j \pmod N}} p^{-1}$$

diverges if and only if the congruence $P_m(x) \equiv j \pmod N$ has a solution $x$ prime to $N$. It suffices now to recall Theorem 5.1 and its Corollary, having in mind, that for prime $p$ one has $m(f,p) = m(f,p^k)$ for $k=1,2,\ldots$ . □

In certain cases one can in the last part of this Corollary replace the assumption $\Lambda_m(f,p^2) = G(p^2)$ by the weaker condition $\Lambda_m(f,p) = G(p)$. One of those cases is presented in the next Proposition.

PROPOSITION 5.8. *Under the assumptions and notation of the preceding corollary, if* $p$ *is a prime larger than* $2n$, *where* $n$ *denotes the degree of the polynomial* $P_m(x)$, *then from* $\Lambda_m(f,p) = G(p)$ *the equality* $\Lambda_m(f,p\ ) = G(p^k)$ *follows for* $k=1,2,\ldots$ .

*Proof*. In view of the previous Corollary it suffices to consider the case $k=2$. Let $a \in G(p^2)$. By assumption one can find $r_1,\ldots,r_s \in G(N)$ satisfying

$$P_m(r_1)\ldots P_m(r_s) \equiv a \pmod p.$$

Observe now, that if the derivative $P'_m(x)$ is divisible by $p$ for all those $x \in G(N)$ for which $P_m(x) \not\equiv 0 \pmod p$, then the polynomial $xP_m(x)P'_m(x)$ has all its values at integer points divisible by $p$. If it does not vanish identically $\pmod p$, then its degree equals at least $p$, and we get $1+n+n-1 = 2n \geq p$ contradicting our assumption. If however it vanishes $\pmod p$ identically, then $P'_m(x) \pmod p$ vanishes identically, thus with a certain polynomial $V(x)$ we have $P_m(x) = V(x^p)$, thus $n\ \ p$, again a contradiction.

Hence we may find $r_0 \in G(N)$ such that $P_m(r_0) \in G(N)$ and $P'_m(r_0) \in G(N)$. Since $P_m(r_0)^{p-1} \equiv 1 \pmod p$, the congruence

$$P_m(r_1)\ldots P_m(r_s)P_m(r_0)^{p-2}P_m(x) - a \equiv 0 \pmod p$$

is solvable (with $x = r_0 \in G(p)$ and $P_m(x) \equiv P_m(r_0) \not\equiv 0 \pmod p$) and since the derivative of the left-hand side does not vanish $\pmod p$ at $x \equiv r_0 \pmod p$ there exists a solution of

$$P_m(r_1)\ldots P_m(r_s)P_m(r_0)^{p-2}P_m(y) \equiv a \pmod{p^2}$$

with $y \equiv r_0 \not\equiv 0 \pmod p$. Since $P_m(r_0), P_m(r_1), \ldots, P_m(r_s) \in \Lambda_m(p^2)$, and $P_m(y) \equiv P_m(r_0) \pmod p$ implies $P_m(y) \in \Lambda_m(p^2)$ we obtain $a \in \Lambda_m(p^2)$, proving $\Lambda_m(p^2) = G(p^2)$. □

## § 3. The number of divisors and Euler φ-function

1. Now we give some examples. Later we shall present an algorithm which for a large class of polynomial-like functions $f$ leads to an effective determination of the set $M^*(f)$, based on a Corollary to Riemann's conjecture for curves proved by A.WEIL, however at this stage we prefer to utilize only simpler tools.

Our first example concerns the number $d(n)$ of positive divisors of $n$. Here the set $M^*(f)$ looks rather complicated. Later we shall learn the reasons for that.

PROPOSITION 5.9. *The function* $d(n)$ *is* WUD(mod N) *if and only if the least prime not dividing* $N$ *is a primitive root* (mod N).

*Proof*. Since $d(p^j) = 1+j$ holds for primes $p$ and $j \geq 1$ the divisor function is polynomial-like. Because $R_j = \{1+j\} \cap G(N)$, the first index $j$ for which it is non-empty equals $q-1$, where $q$ denotes the least prime not dividing $N$, thus $\Lambda_{q-1} = gp\{q\}$ and we see that if $q$ is a primitive root (mod N), then $d(n)$ is WUD(mod N). To prove the converse it suffices to show that if $\Lambda_{q-1} \neq G(N)$ then (5.2) cannot be satisfied with a character $X$ mod $N$ and a prime $p$. To obtain this we assume that (5.2) holds and putting $z = p^{-1(q-1)}$ write

$$0 = \sum_{j=0}^{\infty} X(d(p^j)) z^j = \sum_{j=0}^{\infty} X(1+j) z^j = 1 + z^{-1} \sum_{j=2}^{\infty} X(j) z^j =$$

$$= 1 + z^{-1} \sum_{k=1}^{N-1} X(k) \sum_{\substack{j \geq 2 \\ j \equiv k \pmod N}} z^j = z^{-1}(1 - z^N)^{-1} \sum_{i=1}^{N-1} X(i) z^i .$$

thus

$$\sum_{i=1}^{N-1} X(i) z^i = 0 .$$

This shows that $z$ must be an algebraic integer, but this is certainly not the case, and this contradiction shows that (5.2) cannot hold. □

One sees easily, that all numbers $N$, for which the least prime not dividing $N$ is a primitive root (mod $N$) are included in the following list: $N=4$, $N=2\cdot 3^a$ $(a\geq 1)$, $N=p^a$ ($p$ - an odd prime $a\geq 1$, and 2 is a primitive root (mod $p^a$)) and $N=2p^a$ ($p$ - an odd prime, $a\geq 1$, 3 is a primitive root (mod $p^a$)).

2. Our second example concerns the Euler function $\varphi(n)$. Here the set $M^*(f)$ looks much simpler:

PROPOSITION 5.10. *The set* $M^*(\varphi)$ *consists of all numbers prime to* 6.

*Proof.* If $N$ is even, then $\varphi$ cannot be WUD(mod $N$) since its values for $n\geq 3$ are all even. Let thus $N$ be odd. Since $P_1(x)=x-1$ and $2\in G(N)$ the set $R_1$ contains 1, thus is non-empty and equals

$$\{r:\ 1\leq r\leq N-1,\ (r(r+1),N)=1\}\ .$$

Now we rule out the possibility of (5.2). Indeed, if it holds then necessarily $p=2$, thus

$$0 = 1+\sum_{j=1}^{\infty} X(\varphi(2^j))2^{-j} = 1+\sum_{j=1}^{\infty} X(2)^{j-1}2^{-j} = 1+\frac{1}{2}\left(1-\frac{X(2)}{2}\right)^{-1} =$$

$$= (3-X(2))/(2-X(2)) \neq 0\ ,$$

a contradiction. Thus $N\in M^*(f)$ if and only if the set $R_1$ generates $G(N)$.

If $N$ is divisible by 3, then every element of $R_1$ is congruent to unity (mod 3), thus $R_1$ does not generate $G(N)$. Assume thus that $3\nmid N$. If $N=p_1^{a_1}\cdots p_t^{a_t}$, then $G(N)$ is the product of the groups $G(p_i^{a_i})$ and so we may represent every element of $G(N)$ in the form

$$[y_1,\dots,y_r] \qquad (y_i\in G(p_i^{a_i}))\ .$$

In particular the set $R_1$ equals

$$\{[y_1,\ldots,y_r]:\ y_i \in G(p_i^{a_i}),\ p_i \nmid 1+y_i\} .$$

To prove that it generates $G(N)$ solve, for given $[y_1,\ldots,y_r]$, the congruences

$$2w_i \equiv y_i \pmod{p_i^{a_i}} \qquad (i=1,2,\ldots,t),$$

define

$$v_i = \begin{cases} y_i & \text{if } y_i \not\equiv -1 \pmod{p_i} \\ 2 & \text{otherwise} \end{cases}$$

and

$$z_i = \begin{cases} 1 & \text{if } y_i \not\equiv -1 \pmod{p_i} \\ w_i & \text{otherwise} \end{cases}$$

and observe that $[v_1,\ldots,v\ ]$ and $[z_1,\ldots,z\ ]$ lie in $R_1$ and their product equals $[y_1,\ldots,y\ ]$. □

A similar but more involved reasoning leads to the determination of $M^*(\sigma)$ for $\sigma(n) = \sum_{d|n} d$. It turns out (J.ŚLIWA [73]) that $M^*(\sigma)$ consists of all integers not divisible by 6. We shall deduce this result later (see Proposition 6.9).

## § 4. The vanishing of the sum (5.2)

1. In the previous two examples we get WUD(mod N) exactly in those cases, when the appropriate group $\Lambda_m(f,N)$ coincided with $G(N)$, One can however easily produce examples of multiplicative functions for which with a suitable $N$ we have WUD(mod N) with $\Lambda_m$ being a proper subgroup of $G(N)$. One needs only to satisfy the equality (5.2) for all characters $X$ of the quotient group $G(N)/\Lambda_m(f,N)$. This can be done for the function $\sigma_2(n) = \sum_{d|N} d^2$ and $N = 40$. In fact $\sigma_2$ is polynomial-like, with $P_1(x) = 1+x^2$, $P_2(x) = 1+x^2+x^4$, thus $m(\sigma_2,40) = 2$ $(P_1(x)$

is always even for odd $x$ and $P_2(x) \equiv 3 \pmod{40}$ has $x=1$ for solution). The set $R_2(\sigma_2,40)$ equals $\{3,11\}$ and in view of $3^4 \equiv 1 \pmod{40}$, $11^2 \equiv 1 \pmod{40}$ and $3,3^2,3^3 \not\equiv 11 \pmod{40}$ the group $\Lambda_2$ generated by $R_2$ has 8 elements, thus is of index two in $G(40)$. The only non-trivial character $X \pmod{40}$ which is trivial on $\Lambda_2$ equals $X = X_0 X_8$, where $X_0$ is the principal character (mod 40) and $X_8$ is the only non-principal character (mod 8) which equals unity for $n=3$. Since $\sigma_2(2^k) = 1+2^2+2^4+\ldots+2^{2k} = (2^{2k+2}-1)/3$, $2^{2k+2}-1 \equiv 7 \pmod 8$ for $k \geq 1$, hence $\sigma_2(2^k) \equiv 5 \pmod 8$, so $X_8(\sigma_2(2^k)) = -1$.

Moreover

$$X_0(\sigma_2(2^k)) = \begin{cases} 0 & k \text{ odd} \\ 1 & k \text{ even}, \end{cases}$$

thus finally

$$1 + \sum_{k=1}^{\infty} X(\sigma_2(2^k))2^{-k/2} = 1 - \sum_{\substack{k>0 \\ 2|k}} 2^{-k/2} = 0 ,$$

and WUD(mod 40) follows from the Corollary to Theorem 5.7.

2. The index of $\Lambda$ can be arbitrarily large without spoiling the weak uniform distribution (see exercise 1), however the next Proposition implies that this is not the case under certain additional assumptions on the values $f(p^k)$. This Proposition allows also in many cases to dispose quickly of the possibility of having WUD(mod N) without $\Lambda(f,N) = G(N)$.

PROPOSITION 5.11. *Let* $N \geq 3$, $M \geq 1$ *be integers,* $X$ *a character* (mod N) *of order* $d$ *and* $a_1, a_2, \ldots$ *a sequence of integers with the property that the sequence* $X(a_j) \pmod N$ *is purely periodic, i.e. there is an integer* $T$ *such that*

$$X(a_{j+T}) \equiv X(a_j) \pmod N \qquad (j=1,2,\ldots) .$$

*If for* $k=1,2,\ldots,d-1$ *there exists a prime* $p=p(k)$ *such that*

$$1 + \sum_{j=1}^{\infty} X^k(a_j)p^{-j/M} = 0$$

*then* $p(k) = 2$, *the character* $X$ *is real, i.e.* $d = 2$ *and moreover*

$$X(a_j) = \begin{cases} -1 & \text{if } M \mid j \\ 0 & \text{if } M \nmid j . \end{cases}$$

*Proof*. Let $T$ be a period of $X(a_j) \pmod N$ and assume without restricting the generality that $T$ exceeds $M$. Let $\chi$ be one of the characters $X^k$ $(k=1,2,\ldots,d-1)$ and denote by $d_k$ its order. We have, with $p = p(k)$,

$$0 = 1 + \sum_{j=1}^{\infty} \chi(a_j) p^{-j/M} = 1 + \sum_{j=1}^{T} \chi(a_r) \sum_{\substack{j \geq 1 \\ j \equiv r \pmod T}} p^{-j/M} =$$

$$= 1 + p^{T/M}(p^{T/M} - 1)^{-1} \sum_{r=1}^{T} \chi(a_r) p^{-r/M} .$$

Putting for shortness $y = p^{1/M}$ we can write this equality in the form

$$y^T + \sum_{r=1}^{T-1} \chi(a_r) y^{T-r} + \chi(a_T) - 1 = 0 . \tag{5.9}$$

Since the values of $\chi(a_i)$ are either zero or roots of unity we cannot have $\chi(a_T) = 0$ or $1$ since in that case $y$ would be an algebraic unit which it is not. Thus $\chi(a_T) - 1$ is a non-zero algebraic integer, divisible by $y$. Denoting by $D$ the order of the root of unity $\chi(a_T)$ we see that $D$ divides $d_k$. Moreover, if $K$ denotes the field generated by $y$ and the primitive d-th root of unity and by $N(x)$ the norm from $K$ to $Q$, then $N(y) \mid N(\chi(a_T) - 1)$. If $D$ is not a prime power then $\chi(a_T) - 1$ is a unit, hence $y$ would be also a unit, which is a nonsense. Thus $D$ must be a prime power, say $D = p_0^s$ and since $N(y)$ is a power of $p$ we get with a certain $t$

$$p(k) \mid N(y) \mid N(\chi(a_T) - 1) = p^t \mid D^t \mid d_k^t$$

hence $d_k$ is divisible by $p(k)$.

Now let $q$ be a prime divisor of $d$ and let $\chi = X^{d/q}$. Then $\chi$ is of order $q$, and applying the last observation we get $q = p(d/q)$. We shall now show that $q = 2$ and to do this we assume, a contrario, that $q=p=p(d/q)$ is an odd prime. Note also that since all non-trivial powers of $\chi$ are also of order $q$ thus with the same value for $y$ ( $=p(d/q)^{1/M}$) we have

$$y^T + \sum_{r=1}^{T-1} \chi^k(a_r)y^{T-r} + \chi^k(a_T) - 1 = 0$$

for $k=1,2,\ldots,q-1$.

Adding these equalities we get

$$(q-1)y^T + (q-1)\sum_{\substack{1\le r\le T-1\\ \chi(a_{T-r})=1}} y^r - \sum_{\substack{1\le r\le T-1\\ \chi(a_{T-r})\neq 0,1}} y^r - q = 0 \qquad (5.10)$$

because

$$\sum_{k=1}^{q-1} \chi^k(c) = \begin{cases} q-1 & \text{if } \chi(c)=1 \\ 0 & \text{if } \chi(c)=0 \\ -1 & \text{if } \chi(c)\neq 0,1 . \end{cases}$$

Now observe that we can write $\chi^k(a_T)=\zeta_q^s$ with a primitive $q$-th root of unity $\zeta_q$ and a suitable $1\le s\le q-1$. Since with an appropriate unit $\varepsilon$ we have

$$q = \prod_{j=1}^{q-1} (1-\zeta_q^j) = \varepsilon^{-1}(1-\zeta_q^s)^{q-1}$$

thus

$$\varepsilon\cdot q = (1-\chi^k(a_T))^{q-1} = (y^T + \sum_{r=1}^{T-1} \chi^k(a_{T-r})y^r)^{q-1}$$

Denote by $j$ the minimal index $\ge 1$ such that $\chi^k(a_{T-j})\neq 0$ and obssrve that it is independent of $k$. Now

$$\varepsilon y^M = \varepsilon q = (y^T + \sum_{r=1+j}^{T-1} \chi^k(a_{T-r})y^r + \chi^k(a_{T-j})y^j)^{q-1} = y^{j(q-1)}\cdot A$$

where $A$ is an integer of the field $K$ not divisible by any prime ideal of $K$ which divides $y$, thus

$$M = j(q-1) \geq 2j > j$$

results.

On the other hand from (5.10) we get, using $y^M = q$,

$$0 \equiv y^T + \sum_{\substack{1 \leq r \leq T-1 \\ (a_{T-r},N)=1}} y^r \equiv y^T + y^j (1 + \sum_{\substack{j<r\leq T-1 \\ (a_{T-r},N)=1}} y^{r-j}) \pmod{y^M} .$$

The right-hand side of this expression is divisible by $y^j$ but not by $y^{j+1}$, hence $j \geq M$ results, a contradiction.

We see thus that $q$ cannot be odd, thus $q=2$ is the only prime divisor of $d$, so $d$ must be a power of $2$. It follows that all numbers $d_k$ are powers of $2$ and since $p(k) | d_k$ we obtain $p(k) = 2$ for $k=1,2,\ldots,d-1$.

Consider now the character $\chi = X^{d/2}$ which is of order 2, hence real. Since from (5.9) follows $y | \chi(a_T)-1$ and $\chi(a_T)-1$ is a rational integer, it must be even hence we obtain $\chi(a_T) = -1$, thus

$$y^T + (a_1)y^{T-1} + \ldots + (\chi(a_{T-M})-1)y^M + \ldots + \chi(a_{T-1})y = 0 .$$

Since $y$ is not a unit we must have

$$\chi(a_{T-1}) = \chi(a_{T-2}) = \ldots = \chi(a_{T-M+1}) = 0$$

thus

$$y^{T-M} + \chi(a_1)y^{T-M-1} + \ldots + \chi(a_{T-M}) - 1 = 0$$

which gives $\chi(a_{T-M}) = -1$, and the repetition of our argument leads to $\chi(a_{T-M}) = -1$ and $\chi(a_{T-M-1}) = \ldots = \chi(a_{T-2M+1}) = 0$.

Continuing in this way we arrive finally at

$$\chi(a_j) = \begin{cases} -1 & \text{if } j \text{ is of the form } T-rM \quad (r \in Z) \\ 0 & \text{otherwise} \end{cases}$$

and since the coefficient of $y^{T-M}$ is non-zero we obtain also the divisibility of $T$ by $M$, thus

$$\chi(a_j) = \begin{cases} -1 & \text{if } M \mid j \\ 0 & \text{otherwise} \end{cases}$$

results.

To conclude the proof it remains to prove $d = 2$. Now from last equalities we get $X(a_j) = 0$ for $M \nmid j$, thus

$$0 = 1 + \sum_{j=1}^{\infty} X(a_j) 2^{-j/M} = 1 + \sum_{\substack{j \geq 1 \\ M \mid j}} X(a_j) 2^{-j/M} = 1 + \sum_{k=1}^{\infty} X(a_{kM}) 2^{-k}$$

and this easily implies $X(a_j) = -1$ for $j$ divisible by $M$. If it were $d \neq 2$, then $X^2$ would be a non-principal character, and so we would have

$$0 = 1 + \sum_{j=1}^{\infty} X^2(a_j) 2^{-j/M} = 1 + \sum_{\substack{j \geq 1 \\ M \mid j}}^{\infty} 2^{-j/M} = 2$$

a clear contradiction. Hence $d = 2$ and we are ready. □

COROLLARY: *Let* $N \geq 3$ *be an integer and* $f$ *a multiplicative integer-valued function, for which* $m(f,N) = m$ *is defined. If for every prime* $p \leq 2^m$ *and every character* $X \pmod N$ *the sequence* $X(f(p^k)) \pmod p$ $(k=1,2,\ldots)$ *is periodic, then* $f$ *will be* D-WUD$\pmod N$ *if and only if either* $\Lambda_m(f,N) = G(N)$ *or* $\Lambda_m(f,N)$ *is a subgroup of index* 2 *in* $G(N)$ *and for only non-principal character* $X \pmod N$, *which is trivial on* $\Lambda_m(f,N)$ *one has*

$$X(f(2^k)) = \begin{cases} -1 & \text{if } m\ k \\ 0 & \text{otherwise.} \end{cases}$$

*If* $f$ *is polynomial-like and* $f(p^m) = V(p)$ *with a polynomial* $V(x)$, *then the second possibility can happen only if* $N$ *is even.*

*Proof*. The first assertion will follow from Theorem 5.1 and the last proposition, once we establish that the index of $\Lambda_m(f,N)$ in

$G(N)$ is in case of WUD(mod $N$) at most 2. The last proposition shows that the corresponding factor group is a power of $C_2$, but if it does not equal $C_2$, then we have at least three non-principal characters, say $X_1, X_2$ and $X_1X_2$, all trivial on $\Lambda_m(f,N)$, so they all should satisfy $X(f(2^m)) = -1$, which is obviously not possible.

If $f(p^m) = V(p)$, $N$ is odd and we have WUD(mod $N$) without $\Lambda_m(f,N) = G(N)$, then $(f(2^m),N) = 1$, thus $(2P(2),N) = 1$, i.e. $P(2) \in \Lambda_m(f,N)$ thus for any character $X$ trivial on $\Lambda_m(f,N)$ we have $-1 = X(f(2)) = X(P(2)) = 1$, a contradiction. □

## § 5. The equality $\Lambda_m(N) = G(N)$.

1. In the last section we analyzed the conditions for the equality (5.2) to hold, and we proved certain results which permit in many cases to eliminate this possibility. In such cases, D-WUD(mod $N$) holds, if and only if for an appropriate value of $m$ one has $\Lambda_m(N) = G(N)$. In this section we shall have a closer look at this condition and shall prove that for a large class of multiplicative functions, which includes the polynomial-like functions as well as certain others, including Ramanujan's $\tau$-function the study of the condition $\Lambda_m(N) = G(N)$ can be reduced to the case, when $N$ is a prime power.

The condition, ensuring it, is the following:

*Let* $m$ *be a positive integer. If* $q_1^{a_1}, \dots, q_s^{a_s}$ *are pairwise coprime prime powers and* $N$ *is their product, then the series*

$$\sum_{\substack{p \\ f(p^m) \equiv r \pmod N}} 1/p \tag{5.11}$$

*diverges for a certain* $r$ *if and only if all series*

$$\sum_{\substack{p \\ f(p^m) \equiv r \pmod{q_j^{a_j}}}} 1/p \qquad (j=1,2,\dots,s) \tag{5.12}$$

*diverge.*

It can be also formulated in the following, equivalent, way:

*If* $R(f,q_j^{a_j})$ $(j=1,2,\dots,s)$, $R(f,N)$ *denote the set of all residue classes* $(\bmod q_j^{a_j})$, *resp.* $(\bmod N)$ *for which (5.12) resp.(5.11) diverges, then* $R(f,N)$ *is the direct product of the sets* $R(f,q_i^{a_i})$ $(i=1,2,\dots,s)$, *if we regard* $G(N)$ *as the direct product of the groups* $G(q_j^{a_j})$.

Observe, that this condition, which we shall denote by $(C_m)$, is satisfied by all polynomial-like multiplicative functions of order at least equal to $m$. Ramanujan's $\tau$-function also satisfies this condition for $m=1,2$ however this fact lies much deeper.

2. We prove now

PROPOSITION 5.12. *Let* $N = \prod_{j=1}^{s} q_j^{a_j} \geq 3$ *be an integer and* $f$ *a multiplicative integer-valued function, for which* $m(f,N)=m$ *is defined. Assume further that* $f$ *satisfies the condition* $(C_m)$. *Then the group* $G(N)$ *will be generated by* $R(f,N)$, *except in the case, when there exist characters* $X_j(\bmod q_j^{a_j})$ $(j=1,2,\dots,s)$, *not all principal, and satisfying* $X_j(n)=c_j$ *for* $n \in R(f,q_j^{a_j})$ *with suitable constants* $c_j$ *having their product* $c_1c_2\dots c_s$ *equal to* 1.

*Proof.* The condition $(C_m)$ gives

$$R(f,N) = \prod_{j=1}^{s} R(f,q_j^{a_j})$$

so we may appeal to the following elementary observation:

LEMMA 5.13. *Let* $A_1,\dots,A_n$ *be finite abelian groups and* $A$ *their direct product. Let for* $i=1,2,\dots,n$, $R_i$ *be a non-empty subset of* $A_i$ *and let* $R \subset A$ *be the direct product of the* $R_i$*'s. Then* $R$ *does not generate* $A$ *if and only if there exist characters* $X_i$ *of* $A_i$ $(i=1,2,\dots,n)$, *not all trivial, which are constant on* $R_i$, *being equal there to* $c_i$, *say and* $c_1\dots c_n=1$.

*Proof.* If such characters $X_1,\dots,X_n$ exists, then their product $X_1\dots X_n$ is non-trivial and equals unity on $R$, hence $R$ lies in the kernel of $X_1\dots X_n$, which is a proper subgroup of $A$. Thus in this case $R$ cannot generate $A$. Conversely, if $R$ does not generate $A$, then there is a character $X$ of $A$, which is non-trivial and equals unity on $R$. Define the charcters $X_i$ of $A_i$ for $i=1,2,\dots,n$ by putting $X_i = X_{|A_i}$. Then clearly $X = X_1 \dots X_n$. If now for $i=1,2,\dots,n$ we have $r_i \in A_i$ then

$$\prod_{i=1}^{n} X_i(r_i) = 1 \tag{5.13}$$

holds and if $1 \le j \le n$ and $s_j \in R_j$, then

$$X_j(s_j) \prod_{\substack{1\le i\le n \\ i\ne j}} X_i(r_i) = 1$$

implying $X_j(r_j) = X_j(s_j)$. Thus $X_j$ must be constant on $R_j$ and if it equals to $c_j$ there, then (5.12) implies $c_1 c_2 \dots c_n = 1$. □

To apply this proposition one needs to know, whether for a given prime power $q^a$ there is a non-principal character $(\mathrm{mod}\ q^a)$ which equals unity on the set $R(f,q^a)$. In the next chapter we shall give a procedure to find all prime powers for which this is possible in the case of polynomial-like functions. Here we shall only point out that in the general case it suffices to deal with primes, their squares and the number 8:

LEMMA 5.14. *Let* $q$ *be a prime,* $a \ge 3$ *and let* $R$ *be a non-empty subset of* $G(q^a)$. *If there exists a non-principal character* $X(\mathrm{mod}\ q^a)$ *which is constant on* $R$, *then there exists also a non-principal character* $\chi(\mathrm{mod}\ q^2)$ *if* $q$ *is odd resp.* $(\mathrm{mod}\ 8)$ *if* $q=2$ *which is constant on the reduction of* $R(\mathrm{mod}\ q^2)$, *resp.* $(\mathrm{mod}\ 8)$. *If* $X$ *is trivial on* $R$, *then* $\chi$ *will be trivial on the reduction of* $R$.

*Proof.* It suffices to note that every power of $X$ must be constant on $R$, and there exists always such a power which is a non-principal character $(\mathrm{mod}\ q^2)$ resp. $(\mathrm{mod}\ 8)$. □

## § 6. Ramanujan's τ-function.

1. Before devoting our attention entirely to polynomial-like functions we want to consider Ramanujan's function $\tau(n)$, which is of importance in the theory of modular forms. We recall, that it is defined by the expansion

$$\sum_{n=1}^{\infty} \tau(n)x^n = x \prod_{j=1}^{\infty} (1-x^j)^{24} \qquad (|x|<1)$$

is multiplicative and for $n \geq 1$ and prime $p$ satisfies

$$\tau(p^{n+1}) = \tau(p^n)\tau(p) - p^{11}\tau(p^{n-1}) . \tag{5.13}$$

(See e.g. T.APOSTOL [76]). The values of $\tau(n)$ for certain small values of $n$, used below, are taken from the table given by J.P.SERRE [68], but with a certain pain can also be directly extracted from the above expansion.

We shall need also certain deeper properties of $\tau(n)$ most of them based on P.DELIGNE [69] which we state in the following three lemmas:

LEMMA 5.15. *Let* $M_1, M_2, j_1, j_2$ *be given integers, satisfying* $(j_1, M_1) = 1$ *and let* $A = A(M_1, M_2, j_1, j_2)$ *be the set of all primes* $p$ *satisfying the conditions* $p \equiv j_1 \pmod{M_1}$, $\tau(p) \equiv j_2 \pmod{M_2}$. *Then there exists a non-negative constant* $c(A)$ *and a function* $g(s;A)$ *regular in the half-plane* $\mathrm{Re}\, s \geq 1$ *such that for* $\mathrm{Re}\, s > 1$ *one has*

$$\sum_{p \in A} p^{-s} = c(A) \log \frac{1}{s-1} + g(s;A) .$$

*The constant* $c(A)$ *vanishes if and only if there is a prime power* $q^k$ *dividing* $(M_1, M_2)$ *such that for every prime* $p \neq q$ *satisfying* $p \equiv j_1 \pmod{q^k}$ *we have* $\tau(p) \not\equiv j_2 \pmod{q^k}$. *This can happen only if belongs to the set* $\{2,3,5,7,23,691\}$.

(This lemma is Théorème 11 of J.P.SERRE [72] with the difference that there it is assumed that $M_2$ is prime to $2\cdot 3\cdot 5\cdot 7\cdot 23\cdot 691$ and concluded that $c(A) > 0$. The proof is in both cases the same.)

LEMMA 5.16. *For primes* $p$ *the following congruences hold:*

(i) $\tau(p) \equiv 1+p \pmod 8$ $(p \neq 2)$

(ii) $\tau(p) \equiv p^2+p^3 \pmod{3^2}$ $(p \neq 3)$

(iii) $\tau(p) \equiv p+p^{10} \pmod{5^2}$ $(p \neq 5)$

(iv) $\tau(p) \equiv p+p^4 \pmod 7$ $(p \neq 7)$

*and if* $(p/7) = -1$ *then* $\tau(p) \equiv p+p^{10} \pmod{7^2}$.

(v) *If* $(p/23) = -1$, *then* $\tau(p) \equiv 0 \pmod{23}$,
*if* $(p/23) = +1$ *and* $p$ *is of the form* $x^2+23y^2$, *then*

$$\tau(p) \equiv 1+p^{11} \pmod{23^2}$$

*and if* $(p/23) = +1$, *but* $p$ *is not of the form* $x^2+23y^2$, *then*

$$\tau(p) \equiv -1 \pmod{23}.$$

(vi) $\tau(p) \equiv 1+p^{11} \pmod{691}$.

(All those congruences, save the second in (v) are classical, and listed in J.P.SERRE [68], and H.P.F.SWINNERTON-DYER [73], where the origins of them are described. The second congruence in (v) was found by H.P.F.SWINNERTON-DYER [73].)

Finally we shall need a result of H.P.F.SWINNERTON-DYER [77] which shows that there are no congruences of this type for higher powers of 7, 23 and 691. (There are similar results for other primes, but we shall not need them here.)

LEMMA 5.17. (i) *If* $(a/7) = 1$ *and* $b \equiv a+a^4 \pmod 7$, *then the constant* $c(A)$ *for* $A = A(7^2, 7^2, a, b)$ *is positive.*

(ii) *If* $(a/23) = -1$ *and* $b \equiv 0 \pmod{23}$, *then* $c(A) > 0$ *for* $A = A(23^2, 23^2, a, b)$, *and the same holds in the case* $(a/23) = +1$ *provided* $b \equiv -1 \pmod{23}$.

(iii) *If* $b \equiv 1+a^{11} \pmod{691}$, *then for* $A(691^2, 691^2, a, b)$ *the constant* $c(A)$ *is positive.*

2. Using these lemmas we prove now

THEOREM 5.18 (J.P.SERRE [75]). *The function* $\tau(n)$ *is* WUD(mod N) *if and only if either* N *is odd and not divisible by* 7 *or* N *is even and not divisible neither by* 3 *nor by* 23.

*Proof.* Lemma 5.15 shows that $\tau$ is decent of order $\geq 1$ and satisfies the condition $(C_1)$ from the last section. Consider first the case of $N$ odd and let

$$N = \prod_{i=1}^{s} q_i^{a_i}$$

be its canonical factorization into prime-powers.

To apply Theorem 5.7 it suffices, in view of the last part of Lemma 5.14, to know that for $i=1,2,\dots,s$ there is at least one prime $p \neq q$ with $q_i \nmid \tau(p)$, but this is easy to establish since for $q_i \neq 3$ we may take $p=2$ (because $\tau(2)=-24$) and for $q_i=3$ the prime $p=7$ is good in view of $\tau(7) \neq -16\ 744 \neq 0 \pmod 3$. Hence Theorem 5.7 is applicable and so we may apply the criterion given by Theorem 5.1. Using it we deduce that $\tau$ will be WUD(mod N) if and only if the set $R_N$ of those residue classes (mod N) which are prime to $N$ and contain values of $\tau(p)$ for $p \nmid N$ generates the group $G(N)$. Indeed, otherwise the equality (5.2) would hold with $m=1$, $p=2$ and every non-principal character $X \pmod N$ equal to unity on $R_N$, thus for such a character we would have

$$X(\tau(2^j)) = -1 \qquad (j=1,2,\dots)$$

however $\tau(2) \bmod N$ lies in $R_N$, and thus $X(\tau(2))=1$, a contradiction.

We dispose now quickly of the case $7|N$. Lemma 5.15 (iv) shows that in this case every element of $R_N$ is a quadratic residue (mod 7) thus $R_N$ cannot generate $G(N)$ and we are ready.

To prove that for $N$ odd and not divisible by 7 we get WUD(mod N) we shall first show that for $k=1,2,\dots,t$ the set $R_k$ of residue classes $(\bmod\ q_k^{a_k})$ prime to $q_k$ containing the value of $\tau(p)$ for $p \neq q_k$ generate $G(q_k^{a_k})$. In view of the Corollary to Theorem 5.1 we may assume that $q_k \leq 2$. Lemma 5.15 shows that for $q_k \neq 3,5,23,691$ our assertion is certainly satisfied. For $q_k=3,5$ we get $2 \in R_k$ by Lemma 5.16 (ii), resp. (iii), taking $p \equiv 1 \pmod{q_k^2}$. Because 2 is a primitive root for all powers of 3 and 5 and the same applies to every integer congruent to $2 \pmod{3^2}$, resp. $\pmod{5^2}$ our assertion results. The case $q_k=23$ can be treated in an analogous way, using the Lemmas 5.15 (v), 5.17 (ii) and the fact that the set of those residues in $G(23^2)$ which are congruent either to 2 or to $-1 \pmod{23}$ generates

$G(23^2)$. Finally for $q_k = 691$ we note that $11 \nmid \varphi(691)$ thus every element of $G(691^2)$ is an 11-th power. Using Lemma 5.17 (iii) we can thus find primes $p \neq 691$ such that $\tau(p)$ is congruent (mod $691^2$) to a given number not congruent to unity (mod 691) and this implies that $R_k$ generates $G(691^2)$.

To conclude the proof it suffices, in view of Proposition 5.12, to show that there cannot exist characters $X_k \pmod{q_k^{a_k}}$ $(k=1,2,\ldots,t)$, not all principal, which satisfy

$$X_k(R_k) = c_k \qquad (k=1,2,\ldots,t)$$

and

$$c_1 c_2 \ldots c_t = 1 \, . \tag{5.14}$$

Assume that this holds. Again using Lemma 5.15 one sees easily that for $q_k \neq 3,5,23$ and 691 the corresponding character must be principal. For $q_k \in \{3,5,23,691\}$ we may, due to Lemma 5.14 assume $a_k \leq 2$.

In case $q_k = 3$ the set $R_k$ consists (in view of Lemma 5.16 (ii) of all residues $\equiv 2 \pmod 3$, hence from

$$c = X_k(2) = X_k(8) = X_k(2)^2$$

we get $X_k(2) = \pm 1$. Since 2 is a primitive root (mod 9) this shows that $X_k$ is either principal, or $X_k(m) = (m/3)$.

In case $q_k = 5$ one deduce from Lemma 5.16 (iii) that $R_k$ consists of all residue classes $\equiv 1 \pmod 5$ and $2 \pmod 5$, thus $X_k(2) = c = X_k(1) = 1$ and since 2 is a primitive root (mod $5^2$) $X_k$ must be principal.

For $q_k = 23$, we have already seen that $R_k = \{x \equiv 2 \pmod{23}\} \cup \{x \equiv -1 \pmod{23}\}$, thus $c^2 = X_k((-1)^2) = 1$ i.e. $c = \pm 1$. If $c = 1$, then $X_k$ is principal, since 2 and -1 generate $G(23)$ and $G(23^2)$. If however $c = -1$, then from $25 \equiv 2 \pmod{23}$ we get $X_k(5^2) = X_k(25) = -1$, hence $X_k(5) = \pm i$, thus the order of $X_k$ must be divisible by 4, which is impossible, since 4 divides neither $\varphi(23)$ nor $\varphi(23^2)$.

Finally, if $q_k = 691$, then as we have seen above, $R_k$ contains all elements of $G(691^2)$ resp. $G(691)$ which are not congruent to unity (mod 691) and so the only character constant on all of them is the principal character.

We see that the only non-principal character $X_k$ occurs in case when one of the primes $q_k$, say $q_1$, equals 3, in which case $c_1 = (\frac{2}{3}) = -1$, but then the product $c_1 \dots c_t$ equals $-1$, contradicting (5.14).

The obtained contradiction proves our theorem in the case $2 \nmid N$. Now assume that $N$ is even. Since for prime $p$ we have $\tau(p^2) = \tau(p)^2 - p^{11}$, thus the value $\tau(p^2) \bmod N$ is determined by $p \pmod N$ and $\tau(p) \bmod N$ and we deduce from Lemma 5.15 that $\tau$ is decent of order $\geq 2$ and satisfies the condition $(C_2)$. As in the previous case we write

$$N = \prod_{i=1}^{s} q_i^{a_i}$$

and observe that for $q_i \neq 2, 23$ we have $q \nmid \tau(2^2) = -2^6 \cdot 23$, and since $\tau(5) = 4830$, hence $\tau(5^2) = 4830^2 - 5^{11} \equiv -5^{11} \not\equiv 0 \pmod{2 \cdot 23}$, thus we may apply Theorem 5.7 and the criterion of theorem 5.1.

First we dispose of the case $23 | N$. Lemma 5.16 (iv) shows that all elements of $R_N$ must in that case be congruent to $3 \pmod{23}$, and since 3 is not a primitive root $\pmod{23}$ because of $3^{11} \equiv 1 \pmod{23}$, $R_N$ cannot generate $G(N)$. It remains to rule out the possibility of (5.2) for $m = 2$ (thus $p = 2$ or 3) for characters $X \pmod N$ which are non--principal, but equal to unity on $R_N$. However for $p = 2$ this is impossible, since in view of (5.13) and $\tau(2) = -24 \equiv 0 \pmod 2$ we have $2 | \tau(2^n)$ for all $n \geq 1$, thus due to $2 | N$ the left-hand side of (5.2) equals unity.

Similarly, if $p = 3$, then due to (5.14) and $\tau(3) = 252 \equiv 0 \pmod 2$ we get $\tau(3^{n+1}) \equiv \tau(3^{n-1}) \pmod 2$ and $\tau(3^{2k+1}) \equiv 0 \pmod 2$ and thus the left-hand side of (5.2) is in absolute value at least equal to

$$1 - \sum_{\substack{2|k \\ k>0}} \frac{1}{3^{k/2}} = 1 - \frac{1}{3} \sum_{k=0}^{\infty} 3^{-k} = \frac{1}{2} > 0 .$$

A similar argument shows also that in case $3 | N$ we cannot have WUD(mod N). Indeed, from Lemma 5.16 (ii) and $\tau(p^2) = \tau(p)^2 - p^{11}$ it follows immediately that all elements of $R_N$ are congruent to unity (mod 3), thus $R_N$ cannot generate $G(N)$. Moreover from (5.13) we infer, using $\tau(2) \equiv 0 \pmod 3$ and $\tau(3) \equiv 0 \pmod 3$ that $\tau(3^n)$ is divisible by 3 for all $n \geq 1$ and $\tau(2^n)$ is divisible by 3 for odd $n$ and congruent to unity (mod 3) for even $n$. Since the character $X(m) = (m/3) X_0(m)$ (with $X_0$ being the principal character (mod N)) equals unity on $R_N$, in case of WUD(mod N) it should satisfy (5.2) with

$p=2$ or 3 and $m=2$, however the obtain congruences for $\tau(2^n)$ and $\tau(3^n)$ show that $X(\tau(2^n)) \geq 0$ and $X(\tau(3^n)) = 0$ holds for all $n \geq 1$, ruling thus out the possibility of (5.2).

The remainder of the argument follows the same lines as in the case of odd N. It suffices to show that for $i=1,2,\ldots,s$ the set $R_k$ of residue classes $(\mathrm{mod}\ q_k^2)$ (if $q_k \neq 2$) resp. (mod 8) (if $q_k = 2$) generates $G(q_k^2)$ resp. $G(8)$ and further, that to non-principal character is constant on it. Lemma 5.15 shows that it is enough to do this for $q_k \in \{2,5,7,691\}$.

In case $q_k = 2$ we use Lemma 5.16 (i) and (5.14) to get $R_k = G(8)$. In the same way we obtain in case $q_k = 5$ that $\tau(p^2) \equiv 1 + p^2 + p^{11} \pmod{5^2}$ hence $R_k$ contains all residues $(\mathrm{mod}\ 5^2)$ congruent to 1 or $3 \pmod 5$. Since 3 is a primitive root (mod 25) this shows that $R_k$ generates $G(5^2)$. Moreover $R_k$ contains at least 10 elements, so the only non--principal character trivial on $R_k$ must be the quadratic character $(n/5)$, but $(1/5) = 1 \neq -1 = (3/5)$ and both 1 and 3 lie in $R_k$.

In case $q_k = 7$ Lemma 5.17 (i) implies that for any $a \pmod{7^2}$ satisfying $a \equiv 1,2,4 \pmod 7$ the set $R_k$ will contain every residue class $(\mathrm{mod}\ 7^2)$ which is congruent to $(a^4+a)^2 - a^{11} \pmod 7$ and this shows that all residues congruent to 3,5 and 6 (mod 7) lie in $R_k$. However $\tau(3^2) = -113\,643 \equiv 2 \pmod 7$ and so we have at least $3\cdot 7 + 1 = 22$ elements in $R_k$ and since $G(7^2)$ has $42 < 2\cdot 22$ elements, the only character constant on $R_k$ must be the principal one.

Finally, if $q_k = 691$, Lemma 5.17 (iii) implies that $R_k$ contains every residue class $(\mathrm{mod}\ 691^2)$ which is congruent to $1 + a^{11} + a^{22} \pmod{691}$, provided $691 \nmid a(1+a^{11}+a^{22})$. Since $11 \nmid \varphi(691^2)$ every residue $(\mathrm{mod}\ 691^2)$ is an 11-th power, thus

$$R_k = \{x \ (\mathrm{mod}\ 691^2) : x \equiv 1+y+y^2 \ (\mathrm{mod}\ 691),\ 691 \nmid y(1+y+y^2)\}$$

and we see that $R_k$ contains at least $691(691-3)/2$ elements. This shows that if a non-principal character is constant on $R_k$, it must be the quadratic character $(n/691)$, however in view of

$$\left(\frac{1+1+1^2}{691}\right) = -1\,, \qquad \left(\frac{1+4+4^2}{691}\right) = +1$$

this does not occur. This establishes the Theorem for N even. □

## § 7. Notes and comments

1. The notion of Dirichlet-WUD(mod N) appears first in W.NARKIEWICZ, J.ŚLIWA [76], where also Theorem 5.1 was proved. Proposition 5.5 appears in W.NARKIEWICZ [77], however the criterion resulting from it and Theorem 5.1 can be found in an equivalent form in H.DELANGE [76]. Delange proved further that for a multiplicative integer-valued function f the set of those n's for which f(n) lies in a fixed residue class (mod N) has always a density. The Corollary to Theorem 5.7 was proved in another way (although also via Delange's tauberian theorem) in W.NARKIEWICZ [66], where also Proposition 5.9 and 5.10 appear. A special case of Proposition 5.9 occurs already in L.G.SATHE [45]. Proposition 5.11 was obtained in a special case in W.NARKIEWICZ, F.RAYNER [82] and in the general case in W.NARKIEWICZ [83b].

2. Theorem 5.18 is due to J.P.SERRE [75], who considered only odd values of N, however the general case presents no new problems. The same approach is in principle applicable also to the Fourier coefficients of other modular forms, provided they are integral and multiplicative and the images of the corresponding $\ell$-adic representations are known. In many cases this image was determined by H.P.F.SWINNERTON-DYER [77].

It would be interesting to study UD and WUD for the partition function p(n) and the Fourier coefficients c(n) of the modular invariant j, however the known methods seem completely unadapted to fulfill this task. It is known only that for certain values of N there are infinitely many values of these functions in every residue class (mod N) prime to N. This happens e.g. for all powers of 13 (both for p(n) and c(n)) (A.O.L.ATKIN, J.N.O'BRIEN [67]). M.NEWMAN [60] conjectured that for every N every residue class (mod N) contains infinitely many values of the partition function p(n), and proved this conjecture for $N=5$ and 13. O.KOLBERG [59] proved it for $N=2$ and T.KLØVE [68] for $N=7,17,19,29$ and 31. In T.KLØVE [70] the case $N=121$ was settled. The case $N=7$ was also solved by A.O.L.ATKIN [68]. Cf. D.W.MCLEAN [80] for numerical results.

3. Uniform distribution (mod N) was considered for multiplicative functions by H.DELANGE [77], who gave necessary and sufficient conditions for that in the case of completely multiplicative ($f(mn)=f(m)f(n)$ for all m,n) and strongly multiplicative (i.e. multiplicative and satisfying $f(p^k)=f(p)$ for all primes p and $k\geq 1$) functions. Using them he deduced the existence of infinitely many multiplicative functions which are UD(mod N) for all N.

## Exercises

1. Show that for any $T$ one can find an integer $N$ and a multiplicative function $f$ with well-defined $m(f,N)$ such that the index of $\Lambda_m(f,N)$ is larger than $T$ and $f$ is D-WUD(mod N).

2. Let $f$ be a multiplicative function and $q$ an odd prime such that $M(f,q)=1$. Prove that if $f$ is D-WUD(mod $q^2$) then it is also D-WUD(mod $q^k$) for all $k \geq 1$.

3. Show that in the preceding excercise one cannot replace the assumption $M(f,q)=1$ by the mere existence of $M(f,q)$.

4. (H.DELANGE [77]). Let $f$ be an integer-valued completely multiplicative function, satisfying $f(p) \geq 2$ for all primes $p$ and which is UD(mod N) for all $N$. Prove that $f(n)=n$.

5. Prove that for a multiplicative integer-valued function $f$ the set of all those $n$'s for which $(f(n),N)=1$ has a positive density if and only if $f$ belongs to the class $F_N$.

6. (H.DELANGE [76]). Deduce from Proposition 4.1 a criterion for WUD(mod N) for functions from $F_N$.

7. Use the Corollary to Theorem 5.7 to deduce the vanishing of the mean-value of the Moebius function $\mu(n)$.

8. (O.M.FOMENKO [80], in the case of prime $N$). Determine all those $N$'s for which the function $r_2(n)$, counting the representations of $n$ as a sum of two squares, is WUD(mod N).

# CHAPTER VI

# POLYNOMIAL-LIKE FUNCTIONS

## § 1. Generating $G(N)$ by the set of values of a polynomial

1. The Corollary to Theorem 5.7 shows that in order to check weak uniform distribution (mod N) for a polynomial-like function it is important to have a procedure for checking whether, for a given polynomial $P(x)$ over $Z$, the set of values attained by $P(x)$ at integers $x$ with the property $(xP(x),N) = 1$ does generate the group $G(N)$. We shall now present such a procedure, which is based on a corollary to A.Weil's theorem on the Riemann conjecture for algebraic curves (A.WEIL [48]), stated below:

LEMMA 6.1. (For a proof see e.g. W.SCHMIDT [76], Ch.II, th.2C). *Let* $p$ *be a prime and let* $P(x)$ *be a polynomial over* $Z$ *of degree* $K$. *Further, let* $X$ *be a non-principal character* (mod p) *and denote by* $d$ *its order. If the polynomial* $P$ *does not satisfy the congruence*

$$P(x) \equiv cW^d(x) \pmod p$$

*with a certain constant* $c$ *and a polynomial* $W(x)$, *then one has*

$$\Big|\sum_{x \pmod p} X(P(x))\Big| \le (K-1)p^{\frac{1}{2}} .$$

First let us deduce from this lemma an answer to a question of P. ERDÖS, posed on one of the number-theoretical meetings at Oberwolfach. He asked, whether a "well-behaved" (in a certain sense) multiplicative function will be necessarily WUD(mod p) for all sufficiently large p.

This cannot hold for all multiplicative functions occuring in a natural way, since by Proposition 5.9 the divisor function $d(n)$ does not have this property. However there is a large class of functions for which the answer to Erdös's question is positive.

PROPOSITION 6.2. *Let* $f$ *be a multiplicative, polynomial-like function. Let* $f(p)=P(p)$ *for all primes* $p$ *with a polynomial* $P(x)$ *of degree* $d \geq 1$ *which is not of the form* $cW^k(x)$ *for a certain constant* $c$*, a polynomial* $W(x) \in Z[x]$ *and* $k \geq 2$. *Then* $f$ *is* WUD(mod p) *for all sufficiently large primes* $p$ .

*Proof*. We need a lemma.

LEMMA 6.3. *If* $P(x) \in Z[x]$ *is a polynomial of degree* $d \geq 1$ *which is not of the form* $P(x)=cW^k(x)$ *with a constant* $c$, $k \geq 2$ *and a polynomial* $W(x)$. *Denote by* $D$ *the discriminant of the product of all irreducible factors of* $P(x)$ *and assume that* $p$ *is a prime which does not divide* $D$*, nor the leading coefficient of* $P(x)$. *Then* $P(x)$ *cannot be congruent* (mod p) *to a polynomial* $cW^k(x)$ *with* $c,k,W(x)$ *as above.*

*Proof*. Assume first that $P(x)$ is monic, i.e. its highest coefficient equals unity. Let $P(x)=V_1^{a_1}(x)\dots V_n^{a_n}(x)$ be the factorization of $P$ into irreducible factors over the rationals. Our assumption implies that $(a_1,a_2,\dots,a_n)=1$. Assume now that $p$ is a prime which does not divide $D$ but with a certain constant $c$, $k \geq 2$ and a polynomial $W(x)$ we have $P(x) \equiv cW^k(x) \pmod p$. Let $\bar{P}$, $\bar{W}$ be the images of $P$ resp. $W$ in $GF(p)$, thus

$$\bar{P}(x) = \bar{c}\bar{W}(x)^k$$

holds with a certain non-zero constant $\bar{c}$. Denote by $K$ the splitting field of $P$ and by $Z_K$ the ring of integers of it. Let $P$ be any prime ideal of $Z_K$ containing $p$ and let $\bar{K}=Z_K/P \simeq GF(p^f)$. Since $P(x)$ splits in $K$, the polynomial $\bar{P}(x)$ must split in $\bar{K}$, thus we may write

$$P(x) = \prod_{i=1}^{n} (x-\alpha_i)^{a_i}$$

with $\alpha_1,\dots,\alpha_r \in Z_K$, distinct, and

$$\overline{P}(x) = \prod_{i=1}^{n} (x - \overline{\alpha}_i)^{a_i} = \overline{c}\,\overline{W}(x)^k.$$

with $\overline{\alpha}_i$ being the image of $\alpha_i$ under the canonical map $Z_K \to \overline{K}$. If all $\overline{\alpha}_i$'s were distinct, then $k$ would divide $(a_1,\ldots,a_n) = 1$ hence it exists a pair $\overline{\alpha}_i = \overline{\alpha}_j$ with $i \neq j$. But then $\alpha_i - \alpha_j \in \mathcal{P}$ so $D$ is divisible by $\mathcal{P}$ and since it is a rational integer we get $p|D$, contrary to our assumption.

If $P(x) = Ax^d + \ldots$ is not monic and $P(x) = A\prod_{i=1}^{n}(x - \alpha_i)^{a_i}$ with distrinct $\alpha_1,\alpha_2,\ldots,\alpha_n$ then for $i=1,2,\ldots,n$ we may write $\alpha_i = \beta_i/q$ with suitable algebraic integers $\beta_1,\ldots,\beta_n$ and a rational positive integer $q$, whose all prime factors divide $A$ hence $p \nmid A$. If we now put

$$F(x) = \prod_{i=1}^{n} (x - \beta_i)^{a_i}$$

then $F(x) = q^d A^{-1} P(x/q)$ and $F(x) \in Z[x]$.

If now for certain $c$, $k \geq 2$ and $W(x) \in Z[x]$ we have

$$P(x) \equiv cW^k(x) \pmod p$$

and we define $A'$, $q'$ by $cc' \equiv AA' \equiv 1 \pmod p$, then

$$F(x) \equiv q^d A' c W(xq')^k \pmod p$$

and since $F(x)$ is monic we must have either $F(x) = aV^k(x)$ for a certain constant $a$, $k \geq 2$ and $V(x) \in Z[x]$ hence

$$P(x) = Aq^{-d}F(qx) = aAq^{-d}V^k(qd)$$

contradicting our assumptions, or the discriminant of the polynomial $\prod_{i=1}^{n}(x - \beta_i)$ is divisible by $p$, but in that case the discriminant of $P(x)$ must be also divisible by $p$, which is impossible. □

To prove the proposition it suffices to show that the set

$$R = \{P(x) : (xP(x),p) = 1\}$$

generates $G(p)$ for sufficiently large $p$. Let thus $p$ be larger than all prime factors of the discriminant of the product of all irreducible factors of $P(x)$ and assume further that $p$ does not divide the leading coefficient of $P(x)$. If $R$ does not generate $G(p)$, then there exists a non-principal character $X(\text{mod } p)$, equal to unity on $R$. Thus for all $x$ satisfying $p \nmid xP(x)$ we have $X(P(x)) = 1$, and we obtain from Lemma 6.1

$$(d-1)\sqrt{p} \geq \left|\sum_{x=0}^{p-1} X(P(x)\right| = \left|\#\{x(\text{mod } p): p\nmid xP(x)\} + X(P(0))\right| \geq p-d-2$$

and thus we have only finitely many possibilities for $p$. □

2. Now we prove

THEOREM 6.4. *Let* $P(x)$ *be a polynomial over* $Z$ *of degree* $d \geq 1$, *not of the form* $cW^k(x)$ *(with a constant* $c$, $k \geq 2$ *and* $W(x) \in Z[x]$) *and let* $p^a$ *be a prime power. If there exists a non-principal character* $X(\text{mod } p^a)$, *constant on the set* $R = \{P(x): p\nmid xP(x)\}$ *then either* $p$ *divides the discriminant* $D$ *of the product of irreducible factors of* $P$, *or the leading term* $a_0$ *of* $P$, *or finally* $p$ *does not exceed* $\max\{d^2+2d,\ 3d+2\}$. *If* $X$ *equals unity in* $R$ *then either* $p|Da_0$ *or* $p \leq d^2+2d$.

*Proof.* We may assume that $p$ is odd and moreover in view of Lemma 5.14 $a \leq 2$. Since $R$ contains at least $p^{a-1}(p-1-d)$ elements, our assumptions imply $R \neq \emptyset$. Observe now that $G(p^2) \simeq G(p) \oplus C_p$, (with $C_p$ being cyclic of $p$ elements) the isomorphism given by

$$x(\text{mod } p^2) \to \langle x(\text{mod } p), \hat{x}\rangle$$

where $\hat{x}$ is the unique element of $G(p^2)$ satisfying $\hat{x} \equiv 1(\text{mod } p)$ and $\hat{x}^{p-1} \equiv x^{p-1}(\text{mod } p^2)$. Moreover the character $X$ acts by

$$X(\langle x \text{ mod } p, \hat{x}\rangle) = \psi(x)\chi(\hat{x})$$

where $\psi$ is a character $(\text{mod } p)$, and $\chi$ is a character $(\text{mod } p^2)$ of order dividing $p$. If $\psi$ is non-principal we proceed as in the proof of Proposition 6.2. Indeed, since $\chi^p = 1$ and $\psi^p$ is non-principal, we have

$$X^p(R) = \psi^p(R) = c$$

with a constant $c$, thus by Lemma 6.1 (which we may utilize in view of Lemma 6.3) we obtain, with $R'$ being the reduction of $R \pmod p$,

$$(d-1)\sqrt{p} \geq \left| \sum_{x=0}^{p} X\,(P(x)) \right| \geq \# R' - 1 \geq p - d - 2$$

thus $p < (d+1)^2$. In case $a = 1$ we are ready

Assume now that $a = 2$ and $\psi$ is the principal character. Since for all $x$, $\hat{x}$ lies in the group generated by $1+p \pmod{p^2}$ we can write

$$\hat{x} \equiv (1+p)^{t(x)} \pmod{p^2}$$

with a unique $t(x)$ satisfying $0 \leq t(x) < p$. If $\chi(1+p) = \eta^r$ with $1 \leq r \leq p-1$ and $\eta = \exp\{2\pi i/p\}$, then for all $x$ we have

$$X(x) = \chi(\hat{x}) = \chi(1+p)^{t(x)} = \eta^{rt(x)} .$$

By assumption, for every $x$ satisfying $p \nmid xP(x)$ we have $X(P(x)) = C_1$, with a certain constant $C_1$, (in sequel $C_2, C_3, \ldots$ all will denote constants) thus

$$\eta^{rt(P(x))} = C_1$$

and so

$$rt(P(x)) \equiv C_2 \pmod p$$

holds implying in turn

$$(1+p)^{rt(P(x))} \equiv C_3 \pmod{p^2}$$

i.e.

$$P(x)^r \equiv C_3 \pmod{p^2}$$

and this shows, that for $x$ subject to $p \nmid xP(x)$ the polynomial $P(x)^r$ assumes at most $p-1$ values $\pmod{p^2}$, all distinct $\pmod p$. Denote these values by $c_1,\dots,c_r$ and let $N(c)$ be the number of solutions $x \not\equiv 0 \pmod p$ of the congruence

$$P(x)^r \equiv c \pmod{p^2} .$$

On one hand we have

$$\sum_{j=1}^{r} N(c_j) = \# R \ge p(p-1-d)$$

and on the other hand

$$N(c) \le \#\{x \bmod p:\ p\nmid x, (P(x)^r)' \not\equiv 0 \pmod p, P^r(x) \equiv c \pmod p\} +$$

$$+ p\,\#\{x \bmod p:\ p\nmid x, (P^r(x))' \equiv 0 \pmod p,\ P^r(x) \equiv c \pmod p\} ,$$

thus with

$$S = \#\{x \bmod p;\ p\nmid x, P(x) \not\equiv 0 \pmod p\ \ (P^r(x))' \not\equiv 0 \pmod p\}$$

we get

$$p(p-1-d) \le \sum_{j=1}^{r} N(c_j) \le S + p(p-S-1)$$

Hence $p \ge S \ge p(S-d)$ thus $S-d \le 1$ and $\rho \le 1+d$. We see thus, that $(P^r(x))'$ vanishes for at least $\#R'-1-d \ge p-2d-2$ residue classes $x \pmod p \in R$.

Now $(P^r(x))' = rP^{r-1}(x)P'(x)$, thus $(P^r(x))' \equiv 0 \pmod p$ for $x \in R$ implies $P'(x) \equiv 0 \pmod p$. Since $P'(x)$ is of degree $d-1$ the last congruence has at most $d-1$ solutions, leading to $p-2d-2 \le d-1$ and $p \le 3(d+1)$, or $P'$ vanishes identically $\pmod p$, in which case $P(x) = P_1(x^p)$ holds with a certain polynomial $P_1$, but this implies $p \le d$, thus in any case $p \le 3d+2$.

If $X$ equals unity on $R$ and $3d+2 > d^2+2d$, then $d = 1$ hence $P(x) = ax+b$, $(P^r(x))' = ra(ax+b)^{r-1}$, thus

$$S = \begin{cases} p-1 & \text{if } p \mid b \\ p-2 & \text{if } p \nmid b \end{cases}$$

and from $S \le 1+d = 2$ we get $p \le 4$, i.e. $p = 2$, $3 \le d^2+2d$. This establishes the theorem. □

## § 2. An algorithm

1. Now we can deduce an effective procedure for determining, for a given polynomial $P(x)$ satisfying the conditions of Theorem 6.4, all integers $N$ for which the set

$$R = R(P,N) = \{P(x) \bmod N: (xP(x),N) = 1\}$$

does not generate $G(N)$.

THEOREM 6.5. *If* $P$ *satisfies the conditions of theorem 6.4, then one can effectively determine integers* $K_1, K_2, \ldots, K_s$ *with the property, that* $R(P,N)$ *does not generate* $G(N)$ *if and only if* $N$ *is divisible by at least one of the* $K_i$*'s.*

*Proof.* The algorithm runs as follows:

First determine the set $A$ of all primes $p$ for which there exists a non-principal character $X(\bmod p^a)$, which is constant on $R(P,p^a)$ with a certain $a \le 2$ for $p$ odd resp. $a \le 3$ for $p = 2$. Theorem 6.4 gives an upper bound for the primes $p$ to be investigated.

In the next step consider all products $\prod_{p \in A} p^{a_p}$ with $a_p \le 2$ for $p \ne 2$ and $a_2 \le 3$ and find out, for which of them the set $R(P, \prod_{p \in A} p^{a_p})$ does not generate $G(\prod_{p \in A} p^{a_p})$. Let $\{K_1, K_2, \ldots, K_s\} = B$ be the set of integers obtained in this way. We shall prove that these integers satisfy our assertion.

In one direction this is easy: if for a certain $i$ we have $N \equiv 0 (\bmod K_i)$, then there exists a character $X \ne 1$ of $G(K_i)$ equal to unity on $R(P,K_i)$, hence that character lifted to a character of $G(N)$ will be non-principal and trivial on $R(P,N)$, thus $R(P,N)$ could not generate $G(N)$.

To prove the converse we need a lemma:

LEMMA 6.6. *Let* $N = \prod_p p^{a_p}$ *be an integer and let* $N_1 = \prod_p p^{b_p}$ *where* $b_p = \min(a_p, 2)$ *if* $p \neq 2$ *and* $b_2 = \min(a_2, 3)$. *If* $P(x)$ *is a polynomial such that* $R(P,N)$ *does not generate* $G(N)$, *then* $R(P,N_1)$ *does not generate* $G(N_1)$ *either.*

*Proof.* It suffices to established the following assertion:

*If* $N \neq N_1$ *and* $X \neq 1$ *is a character of* $G(N)$ *trivial on* $R(P,N)$, *then some power of it will be a non-principal character* (mod $N_2$) *and trivial on* $R(P,N_2)$ *for a certain proper divisor* $N_2$ *of* $N$.

Write $X = \prod_p X_p$ where $X_p$ is a character of $G(p^{a_p})$ for $p|N$. The order of $X_p$ is for $p \neq 2$ of the form $d_p p^{t_p}$ with $d_p|p-1$ and $0 \leq t_p \leq a_p - 1$ and for $p=2$ of the form $2^{t_p}$ with $0 \leq t_p \leq a_p - 2$ in case $a_p \geq 3$ and $t_p \leq 1$ in case $a_p \leq 2$. Let $s = \max\{t_p : p|N\}$ and let $A$ be the set of all prime divisors $p$ of $N$ with $t_p = s$. If $s \leq 2$ or $s = 3$ and $A$ consists of the single prime $2$, then $N = N_1$ so let us assume that this is not the case. Denote by $M$ the product of all odd primes in $A$ if $s = 3$ and of all primes in $A$ if $s \geq 4$ and consider the character $X' = X^M = \prod_p X^M$. Since for $p \in A$ the order of $X_p^M$ equals $d_p p^s/(d_p p^s, M) = d'_p p^{s-1}$ (with $d'_p|p-1$) thus $X_p^M$ is induced by a character (mod $p^{a_p - 1}$) and hence $X^M$ can be regarded as a character of $G(N/M)$. Since $X^M$ is non-principal, trivial on $R(P,N/M)$ and $N/M$ is a proper divisor of $N$ we are ready. □

Now assume that $R(P,N)$ does not generate $G(N)$. In view of the last lemma we may assume that $N$ is not divisible neither by $2^4$ nor by a cube of an odd prime. If $X \neq 1$ is a character (mod $N$) trivial on $R(P,N)$, then by proposition 5.12 we have $X = \prod_{p|N} X_p$ where $X_p$ is a character (mod $p^{a_p}$) with $p^{a_p} \| N$ and $X_p$ equals $c_p$ on $R(P,p^{a_p})$ with suitable constants $c_p$ satisfying $\prod_{p|N} c_p = 1$. If $C$ is the set of all primes $p|n$ for which the character $X_p$ is non-principal, then $C \subset A$ and moreover $\prod_{p \in C} X_p$ equals unity on $R(P,\hat{N})$ with $\hat{N} = \prod_{p \in C} p^{a_p}$ proving that $R(P,\hat{N})$ does not generate $G(\hat{N})$. The theorem follows now from the observation that $\hat{N}|N$ and $\hat{N} \in B$. □

Primes $p \in A$ will be called *exceptional* for the polynomial $P$.

2. The following question arises immediately:

PROBLEM IV. *Characterize finite sets* $\{K_1, K_2, \ldots, K_n\}$ *(with none of the* $K_i$*'s divisible by* $2^4$ *or a cube of an odd prime) for which there exists a polynomial* $P(x) \in Z[x]$ *such that* $R(P,N)$ *generates* $G(N)$ *if and only if* $N$ *is not divisible by any of the* $K_i$*'s*

It is clear that some conditions have to be imposed on the $K_i$'s, since e.g. one cannot have $n=1$, $K_1 = (p_1p_2)^2$ with $p_1 < p_2$ being odd primes, as the following argument shows: assume that $P(x)$ is a polynomial with the needed property, and call, for shortness, an integer $N$ "bad" if $R(P,N)$ does not generate $G(N)$. We shall show that the number $p_1^2p_2$ must be "bad". In fact, Proposition 5.12 implies the existence of characters $X_i \pmod{p_i^2}$ $(i=1,2)$ such that

$$X_i(R(P,p_i^2)) = c_i \qquad (i=1,2)$$

and $c_1c_2 = 1$. Now the $c$ 's must be roots of unity of the same order $d \neq 1$, say. Since $d \mid (\varphi(p_1^2), \varphi(p_2^2)) = (p_1(p_1-1), p_2(p_2-1))$ and $p_1 < p_2$, $d$ cannot be divisible by $p_2$, thus $d \mid p_2-1$, showing that $X_2$ is induced by a character $\pmod{p_2}$ and this implies that $p_1^2p_2$ is "bad". (See also exercise 5 which shows that already in case $n=1$ the solution of the problem looks rather complicated.)

## § 3. Applications to the study of $M^*(f)$

1. Our first application is an immediate consequence of Theorem 6.5 and strengthens Proposition 6.2:

PROPOSITION 6.7. *If* $f$ *is an integer-valued multiplicative function such that for primes* $p$ *one has* $f(p) = P(p)$ *with a polynomial* $P(x)$ *not of the form* $cW^k(x)$ *(with* $k \geq 2$*, a constant* $c$ *and a polynomial* $W(x)$*) then there exists an effectively determinable finite set* $E$ *of primes with the property, that if* $N$ *has no prime divisors in* $E$*, then* $f$ *is* WUD(mod $N$).

*Proof.* The assertion results from Theorem 6.5 and the Corollary to Theorem 5.7. □

From Theorem 6.5 and the Corollary to Theorem 5.7 one deduces immediately an algorithm for determining $M^*(f)$ for a polynomial-like multiplicative function $f$, provided it satisfies the following three conditions:

(A) *The value of* $m(f,N)$, *as defined in the Corollary to Theorem 5.7, is for all* $N$ *bounded by a constant* $T$.

(B) *For* $k=1,2,\dots,T$ *one has* $f(p^k)=P_k(p)$ *for all primes* $p$, *with* $P_k(x)\in Z[x]$ *and* $P_k(x)$ *not equal to a constant multiple of a* $d$*-th power of a polynomial with* $d\geq 2$.

(C) *If for an integer* $N$ *the set*

$$R(P_m,N) = \{P_m(x) \bmod N:\ (xP_m(x),N)=1\}$$

*(with* $m=m(f,N)$*) does not generate* $G(N)$, *then* $f$ *cannot be* WUD(mod$N$).

The last condition says that the equality (5.2) is irrelevant for the determination of $M^*(f)$. In practice it is mostly checked with the use of Proposition 5.11, the case of a quadratic character being disposed by a separate argument. For further reference we state now a simple observation which will falicitate this task:

LEMMA 6.8. *If a polynomial-like multiplicative function* $f$ *violates for a certain* $N=\prod_p p^{a_p}$ *the condition* (C) *then it does the same already for the integer* $N_1=\prod_p p^{b_p}$ *with*

$$b_p = \begin{cases} \min(2,a_p) & \text{if } p \text{ is odd} \\ \min(3,a_p) & \text{if } p=2. \end{cases}$$

*Proof*. Observe first that $m(f,N)=m(f,N_1)$, since $N$ and $N_1$ are composed of the same prime factors. If $R(P_m,N)$ does not generate $G(N)$ then Lemma 6.6 shows that $R(P_m,N_1)$ does not generate $G(N_1)$ either and since $f$ is assumed to be WUD(mod $N$), it is also WUD(mod $N_1$), since $N_1|N$ and every prime dividing $N$ is also a prime factor of $N_1$. Thus (C) fails for the number $N_1$. □

2. To illustrate our algorithm we consider now the sum of divisors $\sigma(n)$ and prove:

PROPOSITION 7.9 (J.Śliwa [73]). *For the function* $\sigma(n) = \sum_{d|n} d$ *one has* $M^*(\sigma) = \{N: 6\nmid N\}$.

*Proof*. Since $P_1(x) = x+1$, $P_2(x) = x^2+x+1$ we have

$$m(\sigma,N) = \begin{cases} 1 & \text{if } 2\nmid N \\ 2 & \text{if } 2|N. \end{cases}$$

Assume first that $6|N$ and observe that in this case all elements in the set $R(P_2,N)$ are congruent to unity (mod 3) hence $R(P_2,N)$ cannot generate $G(N)$. In case of WUD(mod N) the condition (5.2) must be satisfied for the character $X(n) = (n/3)X_0(n)$, (where $X_0$ denotes the principal character (mod N)) with $p=2$ or 3. Since $\sigma(2^n) = 2^{n+1}-1 \equiv 0$ or 1 (mod 3), we have $X(\sigma(2^n)) \geq 0$, thus for $p = 2,3$ the equality (5.2) cannot hold, and similarly, since $\sigma(3^n) \equiv 1 \pmod 3$ we get $X(\sigma(3^n)) \geq 0$. Thus for $6|N$ WUD(mod N) is excluded.

Assume now $6/N$. If $N$ is odd, then our algorithm obliges us to check the prime powers $3,3^2,5,5^2$ and all their products. Since $R(P_1,3^2) = \{2,5,8\}$ and $R(P_1,5^2)$ contains all $n \pmod{5^2}$ with $n \not\equiv 0,1 \pmod 5$ we obtain that the only exceptional prime is $p=3$ and this shows that for every odd $N$ we have WUD(mod N).

If $N$ is even, and $3\nmid N$, then we have to check $2^a, 5^b, 7^b$ with $a \leq 3$, $b \leq 2$ and their products a short computation shows that in all cases $R(P_2,N)$ generates $G(N)$. □

In a similar way one can find the set $M^*(\sigma_2)$ for $\sigma_2(n) = \sum_{d|n} d^2$ (W.NARKIEWICZ, F.RAYNER [82]) although here certain additional problems arise with the determination of those N's for which the condition (5.2) is satisfied. We do not enter into the details here, and quote only the final result:

*The function* $\sigma_2(n)$ *is* WUD(mod N) *for all integers* $N$ *except when*

(i) $8|N$, $40\nmid N$,

*or*

(ii) $40|N$ *and there is a prime divisor* $p$ *of* $N$ *such that* $p \geq 7$ *and the order of* $4 \pmod p$ *is odd,*

*or finally*

(iii) $N$ *is divisible by one of the numbers* 12, 15, 28, 42 *or* 66.

3. The above result shows that the set $M^*(f)$ may have a complicated structure even for relatively simple polynomial-like functions. If however $f$ fulfills the conditions (A), (B), (C) then the structure of $M^*(f)$ can be described a priori:

THEOREM 6.10. *Assume that* $f$ *is a polynomial-like multiplicative function, satisfying* (A), (B) *and* (C). *Then there exist integers* $D_1, D_2, \ldots, D_T$ *with* $T = \max_N m(f,N)$ *and finite sets* $X_1, \ldots, X_T$ *of integers such, that if* $S(X)$ *denotes the set of all positive integers without a factor in* $X$, *then*

$$M^*(f) = \{N: (N,D_j) \neq 1 \quad for \quad j \le k-1, \ (N,D_k) = 1, \ N \in S(X_k)\}$$

*All integers* $D_i$ *and sets* $X_i$ *can be effectively determined.*

*Proof.* Let $A_j$ be for $j=1,2,\ldots,T$ the set of all $N$'s for which $m(f,N) = j$, and let $f(p^j) = P_j(p)$ for all primes $p$ with $P_j(x) \in Z[x]$. For $N \in A_j$ the function $f$ will be WUD(mod $N$) if and only if $R(P_j,N)$ generates $G(N)$. Due to Theorem 6.5 this happens if and only if $N \in S(X_j)$ where $X_j$ is the set of integers corresponding to $P_j$ by Theorem 6.5.

Observe now that $N \in A_j$ if and only if the sets $R(P_i,N)$ are empty for $i=1,2,\ldots,j-1$ but $R(P_j,N)$ is non-empty and note that to a given polynomial $P_j(x)$ there corresponds an integer $D_j$ with the property that $R(P_j,N) \neq \emptyset$ if and only if $(D_j,N) = 1$. Indeed, if $N = \prod_p p^{a_p}$ then $R(P_j,N) = \prod_p R(P,p^{a_p})$ and since $R(P_j,p^{a_p}) \neq 0$ is equivalent to $R(P_j,p) \neq 0$ and the set $R(P_j,p)$ can be empty only for finitely many primes $p$ (by Theorem 6.5), hence it suffices to put $D_j = \prod p$, the product taken over all primes $p$ for which $R(P_j,p) = \emptyset$. □

## § 4. The functions $\sigma_k$ for $k \ge 3$

1. Our last section is devoted to the functions $\sigma_k(n) = \sum_{d|n} d^k$, with which we dealt already in the cases $k=0,1,2$ and which for odd $k$ are the Fourier coefficients of Eisenstein series. We shall show that the algorithm of § 2 is applicable in this case and using it we determine

the set of all odd N's for which $\sigma_k(n)$ is WUD(mod N) in the case, when $k$ is an odd prime. We start with

PROPOSITION 6.11. *If* $k \geq 3$, *then* $\sigma_k$ *satisfies the conditions* (A), (B), (C) *of the preceding section, hence there is an effective procedure for determining* $M^*(\sigma_k)$.

*Proof.* We have in this case

$$P_j(x) = 1 + x^k + x^{2k} + \ldots + x^{jk} = (x^{k(j+1)} - 1)/(x^k - 1)$$

for $j \geq 1$, thus (B) is obviously satisfied. To prove (A) we shall establish, that if $\pi$ is the smallest prime number satisfying $\pi - 1 \nmid k$, then $m(\sigma_k, N) \leq \pi$. Indeed, if $R(P_{\pi-1}, N)$ is empty, then for a certain prime $p$ dividing $N$ the set $R(P_{\pi-1}, p)$ must be empty, and this implies

$$p \mid 1 + 1 + 1^2 + \ldots + 1^{(\pi-1)} = \pi$$

thus $p = \pi$ and moreover for a primitive root $g \pmod p$ we would have

$$p \mid (g^{kp} - 1)/(g^k - 1)$$

thus $g^{kp} \equiv 1 \pmod p$ but this implies $p-1 \mid kp$, hence $\pi - 1 = p - 1$ must divide $k$, but this is excluded by our choice of $\pi$. (One can show that in fact $\max_N m(\sigma_k, N)$ equals $\pi - 1$).

It remains to show that the condition (C) holds. Here we shall use the Corollary to Proposition 5.11 and to be able to do this we prove first an auxiliary result:

LEMMA 6.12. *If* $N$ *is not divisible by a fourth power of a prime and* $k \geq 3$, *then for every prime* $q$ *the sequence* $\sigma_k(q^j) \pmod N$ $(j=1,2,\ldots)$ *is periodic.*

*Proof.* It suffices to consider $N = p^a$ with $p$ prime and $a \leq 3$. Since for $j, T \geq 1$ we have

$$\sigma_k(q^{j+T}) - \sigma_k(q^j) = q^{k(j+1)}(q^{kT} - 1)/(q^k - 1),$$

hence, if $p \neq q$ and $b$ is defined by $p^b \| q^k - 1$ then the order of

$q^k \pmod{p^{a+b}}$ is a period of $\sigma_k(q^j) \pmod{p^a}$. If however $p = q$, then in view of $a \le 3$ $k(j+1)$ we get

$$p^a \mid \sigma_k(q^{j+T}) - \sigma_k(q^j)$$

thus our sequence is constant $\pmod{p^a}$. □

Since Lemma 6.8 permits as to consider only those N's which are not divisible neither by $2^4$ nor by a cube of an odd prime the above lemma and the corollary to Proposition 5.11 imply now that if $\sigma_k$ is WUD(mod N) but the set $R(P_m,N)$ (with $m = m(\sigma_k,N)$) does not generate $G(N)$ then it generates a subgroup of index two and for the only non--principal character $X \pmod N$ which is trivial on $R(P_m,N)$ one has

$$X(\sigma_k(2^j)) = X\left(\frac{2^{k(j+1)}-1}{2^k-1}\right) = \begin{cases} -1 & \text{if } m \mid j \\ 0 & \text{if } m \nmid j . \end{cases}$$

Moreover in that case N must be even and X real.

Since $k \ge 3$, we have $\sigma_k(2^j) \equiv 1 \pmod 8$ for $j \ge 1$, thus if we write $X = \prod_{p \mid N} X_p$, where $X_p$ is a character $\pmod{p^{a_p}}$ (with $p^{a_p} \| N$) then $X_2(\sigma_k(2^j)) = 1$ holds for $j \ge 1$. If p is a prime dividing N, but not dividing $2^k - 1$ and $r_p$ denotes the order of $2^k \pmod{p^{a_p}}$, then for j divisible by

$$D = \prod_{\substack{p \mid N \\ p \ne 2 \\ p \nmid 2^k-1}} r_p$$

we get $\sigma_k(2^j)$ $1 \pmod{p^{a_p}}$, thus $X_p(\sigma_k(2^j)) = 1$ holds.

It follows that if

$$X' = \prod_{\substack{p \mid N \\ p \ne 2 \\ p \mid 2^k-1}} X_p$$

then for $j \equiv 0 \pmod D$ we have

$$X'(\sigma_k(2^j)) = \begin{cases} -1 & \text{if } m \mid j \\ 0 & \text{if } m \nmid j . \end{cases}$$

Since $X$ was real, $X'$ is also such, hence for $p \neq 2$ the character $X_p$ is induced by a character (mod p). But for $p | 2^k - 1$ we have $\sigma_k(2^j) \equiv 1 + j \pmod p$ thus for such primes $p$ we have $X_p(\sigma_k(2^j)) = X_p(1+j)$, thus $X'(\sigma_k(2^j)) = X'(1+j)$ and this leads to a contradictory equality

$$1 = X'(1 + Dm)X'(1 + Dm) = X'(1 + Dm(2 + Dm)) = -1 . \quad \square$$

2. Utilizing this result F.RAYNER [83] has computed the sets $M^*(\sigma_k)$ for all odd $k \le 107$. For very small $k$ one can do this by hand, as shown in W.NARKIEWICZ [83b] in the case $k=3$. We shall not go into the details of the computation, and prove instead a partial result of E.DOBROWOLSKI (oral communication), which describes the odd numbers belonging to $M^*(\sigma_k)$ in the case of prime $k \ge 3$. Earlier O.M.FOMENKO [80] proved that for any $k$ the set $M^*(\sigma_k)$ cannot contain a prime number exceeding $(k+1)^2$. This can be also deduced from Theorem 6.4.

THEOREM 6.12. *Let* $k$ *be an odd prime and denote by* $M_2^*(\sigma_k)$ *the set of all those odd integers* $N$ *for which* $\sigma_k$ *is* WUD(mod N). *If* $2k + 1$ *is composed, then* $M_2^*(\sigma_k)$ *consists of all odd integers. If however* $q = 2k + 1$ *is a prime, then* $M_2^*(\sigma_k)$ *consists of all integers not divisible by* $K$ *where*

$$K = \begin{cases} q & \text{if } k \equiv 3 \pmod 4 \\ 3q & \text{if } k \equiv 1 \pmod 4 . \end{cases}$$

*Proof*. In our case $P_1(x) = x^k + 1$ and since for odd $N$ the set $R(P_1, N)$ contains $2 = 1 + 1^k$, it is non-empty, thus $m(\sigma_k, N) = 1$. Proposition 6.11 shows now that $\sigma_k$ will be WUD(mod N) if and only if $R(P_1, N)$ generates $G(N)$. First we shall show the only exceptional primes are $p=3$ and $p = q = 2k+1$ provided the last number is a prime. To do this assume that $p$ is an odd prime $a \le 2$ and $X$ a character of $G(p^a)$ which assumes a constant value on $R(P_1, p^a)$. If $H$ denotes the kernel of $X$ then $H$ is a proper subgroup of $G(p^a)$, $R(P_1, p^a)$ must be contained in a coset of $G(p^a)$ with respect to $H$, and since $p$ is an odd prime, thus $G(p^a)$ is cyclic and $H$ coincides with the set of all $r$-th powers of elements of $G(p^a)$ with a certain $r$.

Thus we see that all elements of $R(P_1, p^a)$ are of the form $cy^r$ with a certain fixed elements $c$ of $G(p^a)$ and suitable $y \in G(p^a)$. Without restricting the generality we may assume that $r$ is a prime.

Now let $g$ be a primitive root (mod $p^a$). If $1+g^k$ is divisible by $p$, then $k$ is divisible by $(p-1)/2$, thus either $p=3$, or $(p-1)/2=k$, since $k$ is a prime. In the first case we find $R(P_1,3)=\{2\}$ and $R(P_1,3^2)=\{2,5,8\}$ thus the quadratic character $X_3(n)=(n/3)$ assumes the value $-1$ on those two sets, and in the second case $p=2k+1=q$ is a prime. In this case

$$1+x^k = 1+x^{(p-1)/2} \equiv 1+(\frac{x}{p})$$

hence $R(P_1,p)=\{2\}$ and thus every character (mod p) is constant on $R(P_1,p)$. Since the derivative of $1+x^{(p-1)/2}$ is not divisible by $p$ for $x \neq 0$ the congruences

$$1+x^{(p-1)/2} \equiv 2+jp \pmod{p^2}$$

have solutions for $j=0,1,\ldots,p-1$ thus $R(P_1,p^2)$ equals $\{2+jp: j=0,1,\ldots,p-1\}$ and every character (mod p) lifted to $G(p^2)$ is constant on it.

This shows that $3$ and $q=2k+1$, in case it is a prime, are certainly exceptional. To prove that there are no others it remains to consider the case, when for no primitive root $g \pmod{p^a}$ $1+g^k$ is divisible by $p$. Then with suitable $y_1,y_2 \not\equiv 0 \pmod{p}$ we have $1+g^k \equiv cy_1^r \pmod{p^a}$, $1+g^{-k} = cy_2^r \pmod{p^a}$ hence

$$cy_1^r \equiv 1+g^k \equiv g^k(1+g^{-k}) \equiv cg^k y_2^r \pmod{p^a}$$

showing that $g^k$ is a $r$-th power in $G(p^a)$, hence $r|k$ and so $r=k$. Moreover with a suitable $y_3,y_4 \not\equiv 0 \pmod{p}$ we have $1-g^k \equiv 1+(-g)^k \equiv cy_3^k$ and $1-g^{2k} \equiv 1+(-g^2)^k \equiv y_4 \equiv cy_4^k \pmod{p^a}$ which leads to

$$cy_4^k \equiv 1-g^{2k} \equiv (1-g^k)(1+g^k) \equiv c^2(y_1y_3)^k \pmod{p^a}$$

and this implies that $c$ is itself a $k$-th power in $G(p^a)$, so we may take $c=1$ and thus every element of $R(P_1,p^a)$ must be a $k$-th power in $G(p^a)$. Since $R(P_1,p^a)$ consists of all elements $1+x^k \pmod{p^a}$ not congruent to $0$ and $1 \pmod{p}$ we obtain easily that $j \pmod{p}$ is a $k$-th power (mod $p^a$) for all $j \not\equiv 0,1 \pmod{p}$ which is possible only if $(k,\varphi(p))=1$, but this means that $H$ is not a proper subgroup of $G(p^a)$. Thus indeed, in this case there are no exceptional primes.

Now we check the products of exceptional primes and their powers. If $2k+1$ is composed, they are $3$ and $3^2$ but from the form of $R(P_1,N)$ in that case it is clear that $G(N)$ is generated by it. Hence we get WUD(mod N) for all odd $N$. If $q=2k+1$ is prime, we have two cases to consider. If $k \not\equiv 3 \pmod 4$, then $(2/q)=+1$, so $2$ cannot be a primitive root $\pmod q$, hence $R(P_1,q)$ does not generate $G(q)$, thus for no multiple of $q$ we can get weak uniform distribution but for other integers we have WUD. If $k \equiv 3 \pmod 4$, then $2$ is a primitive root $\pmod q$ hence for $N=q$ and $q^2$ the group $G(N)$ is generated by $R(P_1,N)$ and the same holds for $N=3$ and $3^2$. However for $N=3q$ we have $R(P_1,N) = \{2\}$ and it cannot generate $G(N)$ since this group is non-cyclic. Hence for multiples of $3q$ we do not get WUD(mod N), and only for them. □

Unfortunately this elementary approach does not seem to work for even $N$'s or for $\sigma_k$ with $k$ composed. The results of computing (F.RAYNER [83]) do not indicate any pattern in the structure of $M^*(\sigma_k)$, except that it obeys theorem 6.10. However in remarkably many cases (19 out of 54) we have $M^*(\sigma_k) = M^*(\sigma) = \{N: 6 \nmid N\}$. It would be interesting to have a solution to the following

PROBLEM V. *Characterize those odd integers* $k$, *for which* $M^*(\sigma_k) = \{N: 6 \nmid N\}$.

## § 5. Notes and comments

1. Theorem 6.4 was proved in W.NARKIEWICZ [82] in a more general form, covering also the case of systems of several polynomials. In W. NARKIEWICZ [81] it was utilized to obtain the algorithm given by Theorem 6.5. As an application of this algorithm all pairs M,N of integers were determined with respect to which the pair $\langle d(n), \varphi(n)\rangle$ is WUD.

2. Proposition 7.9 was originally proved by J.ŚLIWA [73] on the basis of the corollary to Proposition 5.7. His proof was longer than that given here, but it did not utilize the deep result of A.Weil, on which Theorem 6.4 rests. In the determination of $M^*(\sigma_2)$ (W.NARKIEWICZ, F.RAYNER [82]) an important role was played by a computer. In fact, since the algorithm of Theorem 6.5 is inapplicable here, a computer

experiment was made, which allowed to formulate a conjecture about the form of integers in $M^*(\sigma_2)$ which was subsequently proved.

Theorem 6.10 occurs in W.NARKIEWICZ [83b], where also Proposition 6.11 was obtained.

3. If $f$ is a polynomial-like multiplicative function which is WUD(mod N) and $N_1$ is a divisor of $N$ which has the same prime factors, then obviously $f$ is also WUD(mod $N_1$). E.J.SCOURFIELD [74] found conditions under which the converse implication holds. She proved namely, that if $m = m(f,N)$ is defined, the polynomial $P_m(x)$ (such that $P_m(p) = f(p^m)$ holds for primes) is non-constant and for a prime $p$ we define

$$\beta_p = \inf\{c \ge 0:\ \text{for a certain } x \text{ with } p \nmid xP_m(x) \text{ one has } p^c \| P_m'(x)\}$$

and put $N_2 = \prod_{p|N} p^{2\beta_p+1}$, then $f$ is WUD(mod N) if and only if it is WUD(mod $N_2$).

## Exercises

1. Determine the set $M^*(J_2)$ for Jordan's function $J_2(n) = n^2 \prod_{p|n} (1-p^{-2})$.

2. Do the same for the function $d_3(n)$, which counts the number of representation of $n$ as a product of three positive factors.

3. Prove Proposition 7.9 without the use of Theorem 6.4.

4. Determine $M^*(\sigma_4)$.

5. Prove that a polynomial $P(x)$ with the property that $R(P,N)$ generates $G(N)$ if and only if $K_1 \nmid N$ exists, if and only if one of the following conditions holds.

(a) $K_1$ is even and has the form $K_1 = 2^\alpha q$ with $1 \le \alpha \le 3$ and $q$ either equal to unity or to an odd prime,

(b) $K_1$ is odd, and has the form $K_1 = p_1 \ldots p_r$ with distinct primes $p_1, \ldots, p_r$ and there exists a common divisor $\delta$ of $p_1-1, p_2-1, \ldots, p_r-1$ stisfying $\delta \ge r$,

(c) $K_1$ is odd and has the form $K_1 = p_0^2 p_1 \ldots p_r$ with distinct primes $p_0, p_1, \ldots, p_r$ and there exists a common divisor $\delta$ of $p_1-1, p_2-1, \ldots, p_{r-1}$ which satisfies $p_0 | \delta | p_0(p_0-1)$ and $\delta \ge 1+r$.

## REFERENCES


APOSTOL, T.M.
[76] Modular functions and Dirichlet series in number theory, Springer 1976. MR 54 # 10149

ATKIN, A.O.L.
[68] Multiplicative congruence properties and density problems for p(n), Proc.London Math.Soc., 18, 1968, 563-576. MR 37 # 2690

ATKIN, A.O.L., O'BRIEN, J.N.
[67] Some properties of p(n) and c(n) modulo powers of 13, Trans. Amer.Math.Soc., 126, 1967, 442-459. MR 35 # 5390

BETTI, E.
[51] Sopra la risolubilitâ per radicali delle equazioni algebriche irredutibili di grado primo, Annali di Sci.Mat.e Fis., 2, 1851, 5-19.

[52] Sulla risoluzioni delle equazioni algebriche, ibidem 3, 1852, 49-115.

[55] Sopra la teorica delle sostituzioni, ibidem, 6, 1855, 5-34.

BRAWLEY, J.V.
[76] Polynomials over a ring that permute the matrices over that ring, J.of Algebra, 38, 1976, 93-99. MR 52 # 13755.

BRAWLEY, J.V., CARLITZ, L., LEVINE, J.
[75] Scalar polynomial functions on the n×n matrices over a finite field, Linear Algebra and Appl., 10, 1975, 199-211. MR 51 # 12800

BRUCKNER, G.
[70] Fibonacci sequence modulo a prime $p \equiv 3 \pmod 4$, Fibonacci Quart., 8, 1970, 217-220. MR 41 # 3384

BUMBY, R.T.
[75] A distribution property for linear recurrence of the second order, Proc.Amer.Math.Soc., 50, 1975, 101-106. MR 51 # 5475

BUNDSCHUH, P.
[74] On the distribution of Fibonacci numbers, Tankang J.Math., 5, 1974, 75-79. MR 50 # 12933

BUNDSCHUH, P., SHIUE, J.S.
[73] Solution of a problem on the uniform distribution of integers, Atti Accad.Lincei, 55, 1973, 172-177. MR 51 # 3072

[74] A generalization of a paper by D.D.Wall, ibidem, 56, 1974, 135-144. MR 51 # 5476

BURKE, J.R., KUIPERS, L.
[76] Asymptotic distribution and independence of sequences of Gaussian integers, Simon Stevin 50, 1976-7, 3-21. MR 54 # 5131

CARLITZ, L.
[53] Permutations in a finite field, Proc.Amer.Math.Soc., 4, 1953, 538. MR 15 p.3

[62a] Some theorems on permutation polynomials, Bull.Amer.Math.Soc., 68, 1962, 120-122. MR 25 # 5052

[62b] A note on permutation functions over a finite field, Duke Math.J., 29, 1962, 325-332. MR 25 # 1151

[63] Permutations in finite fields, Acta Sci.Math.(Szeged), 24, 1963, 196-203. MR 28 # 81.

CARLITZ, L., LUTZ, J.A.
[78] A characterization of permutation polynomials over a finite field, American Math.Monthly 85, 1978, 746-748. MR 80a # 12022

CARMICHAEL, R.D.
[20] On sequences of integers defined by recurrence relations, Quart.J.Math. (Oxford), 48, 1920, 343-372.

CAVIOR, S.R.
[63] A note on octic permutation polynomials, Math.Comput., 17, 1963, 450-452. MR 27 # 3628

CHAUVINEAU, J.
[65] Complément au théorème métrique de Koksma dans R et dans $Q_p$, C.R.Acad.Sci. Paris, 260, 1965, 6252-6255. MR 31 # 3400

[68] Sur la répartition dans R et dans $Q_p$, Acta Arith., 14, 1968, 225-313. MR 39 # 3865

COHEN. S.D.
[70] The distribution of polynomials over a field, Acta Arith., 17, 1970, 255-271. MR 43 # 3234

DABOUSSI, H., DELANGE, H.,
[82] On multiplicative arithmetical functions whose modulus does not exceed one, J.London Math.Soc., 26, 1982, 245-264.

DAVENPORT, H., LEWIS, D.J.
[63] Notes on congruences, I, Quart.J.Math. (Oxford), 14, 1963, 51-60. MR 26 # 3657.

DEDEKIND, R.
[97] II Supplement to Dirichlet's "Vorlesungen über Zahlentheorie", Braunschweig 1897.

DELANGE, H..
[54] Généralisation du théorème de Ikehara, Ann.Scient.Èc.Norm.Sup., 71, 1954, 213-242. MR 16 p.921

[56] Sur la distribution des entiers ayant certaines propriétés, ibidem, 73, 1956, 15-74. MR 18 p.720

[61] Un théorème sur les fonctions multiplicatives et ses applications, ibidem, 78, 1961, 1-29. MR 30 # 71

[69] On integral-valued additive functions, J.Number Th., 1, 1969, 419-430. MR 40 # 1359.

[72] Sur la distribution des valeurs des fonctions additives, C.R. Acad.Sci. Paris, 275, 1972, A 1139-A 1142. MR 46 # 7187

[74] On integral-valued additive functions, II, J.Number Th., 6, 1974, 161-170. MR 49 # 7227

[76] Sur les fonctions multiplicatives à valeurs entiers, C.R.Acad. Sci. Paris, 283, 1976, A 1065-A 1067. MR 55 # 300

[77] ———., ibidem 284, 1977, A 1325-A 1327. MR 56 # 290

DELIGNE, P.
[69] Formes modulaires et représentations $\ell$-adiques, Séminaire Bourbaki, 1969, nr.355.

DESHOUILLERS, J.M.

[73] Sur la répartition des nombres $[n^c]$ dans les progressions arithmétiques, C.R.Acad.Sci. Paris, 227, 1973, A 647-A 650. MR 49 #2603.

DICKSON, L.E.

[97] The analytic representation of substitutions on a power of a prime number of letters with a discussion of the linear group, Ph.D.Thesis, Chicago 1897 = Annals of Math., 11, 1896-97, 65-120 and 161-183.

[01] Linear Groups, Leipzig 1901.

DIJKSMA, A., MEIJER, H.G.

[69] Note on uniformly distributed sequences of integers, Nieuw Arch.Wisk., 17, 1969, 210-213. MR 41 # 1676.

DOWIDAR, A.F.

[72] Summability methods and distribution of sequences of integers, J.Nat.Sci.Math., 12, 1972, 337-341. MR 50 # 2098.

ENGSTROM, H.T.

[31] On sequences defined by linear recurrence relations, Trans. Amer.Math.Soc., 33, 1931, 210-218.

FOMENKO, O.M.

[80] The distribution of values of multiplicative functions with respect to a prime modulus, Zapiski Naučn.Sem. LOMI, 93, 1980, 218-224 (in Russian). MR 81k # 10068

FRIED, M.

[70] On a conjecture of Schur, Michigan Math.J., 17, 1970, 41-55. MR 41 # 6188.

FRYER, K.D.

[55] Note on permutations in a finite field, Proc.Amer.Math.Soc., 6, 1955, 1-2. MR 16 p.678

HALL, M.

[38a] An isomorphism between linear recurring sequences and algebraic rings, Trans.Amer.Math.Soc., 44, 1938; 196-218.

[38b] Equidistribution of residues in sequences, Duke Math.J., 4, 1938, 691-695.

HAYES, D.R.,

[67] A geometric approach to permutation polynomials over a finite field, Duke Math.J., 34, 1967, 293-305. MR 35 # 168.

HERMITE, E.

[63] Sur les fonctions de sept lettres, C.R.Acad.Sci. Paris, 57, 1863, 750-757 = Oeuvres, II, 280-288.

HULE, H. MÜLLER, W.B.

[73] Cyclic groups of permutations induced by polynomials over Galois fields. An. Acad.Brasil., 45, 1973, 63-67. MR 48 # 8453 . (In Spanish).

KLØVE, T.

[68] Recurrence formulae for the coefficients of modular forms and congruences for the partition function and for the coefficients of $j(\tau)$, $(j(\tau)-1728)^{1/2}$ and $j(\tau)^{1/3}$, Math.Scand., 23, 1968, 133-159. MR 40 # 5545.

[70] Density problems for p(n), J.London Math.Soc., 2, 1970, 504--508. MR 42 # 219.

KNIGHT, M.J., WEBB, W.A.

[80] Uniform distribution of third-order linear sequences, Acta Arith., 36, 1980, 17-20.

KOKSMA, J.
[35] Ein mengentheoretischer Satz über die Gleichverteilung modulo Eins, Compos.Math., 2, 1935, 250-258.

KOLBERG, O.
[59] Note on the parity of the partition function, Math.Scand., 7, 1959, 377-378. MR 22 # 7995

• KRONECKER, L.
[81] Zur Theorie der Elimination einer Variabeln aus zwei algebraischen Gleichungen, Monatsber. Kgl. Preuss. Akad. Wiss. Berlin 1881, 535-600 = Werke, II, 113-192.

KUIPERS, L.
[79] Einige Bemerkungen zu einer Arbeit von G.J.Rieger, Elem. Math., 34, 1979, 32-34. MR 80h # 10063

KUIPERS, L., NIEDERREITER, H.
[74a] Asymptotic distribution (mod m) and independence of sequence of integers, I, II, Proc. Japan Acad. Sci., 50, 1974, 256-260 and 261--265. MR 51 # 404

[74b] Uniform distribution of sequences, Wiley-Interscience 1974. MR 54 # 7415

KUIPERS, L., NIEDERREITER, H., SHIUE, J.S.
[75] Uniform distribution of sequences in the ring of Gaussian integers, Bull. Inst. Math. Sin., 3, 1975, 311-325. MR 54 # 2612.

KUIPERS, L., SHIUE, J.S.
[71] On the distribution modulo M of generalized Fibonacci numbers, Tamkang J. Math., 2, 1971, 181-186. MR 46 # 5231

[72a] A distribution property of the sequence of Fibonacci numbers, Fibonacci Quart., 10, 1972, 375-376, 392. MR 47 # 3302

[72b] A distribution property of the sequence of Lucas numbers, Elem. Math., 27, 1972, 10-11. MR 46 # 144

[72c] Asymptotic distribution modulo m of sequences of integers and the notion of independence, Atti Accad. Lincei Mem. Cl. Sci. Fis. Mat. Nat., 11, 1972-3, 63-90. MR 51 # 3106

[72d] A distribution property of a linear recurrence of the second order, Atti Accad. Lincei, 52, 1972, 6-10. MR 48 # 3862.

[80] On a criterion for uniform distribution of a sequence in the ring of Gaussian integers, Rev. Roum. Math. Pures Appl., 25, 1980, 1059-1063. MR 81m # 10070

KUIPERS, L., UCHIYAMA, S.
[68] Note on the uniform distribution of sequences of integers, Proc. Japan Acad., 44, 1968, 608-613. MR 39 # 148.

KURBATOV, V.A.
[49] On the monodromy group of an algebraic function, Mat. Sbornik, 25, 1949, 51-94. MR 11 p.85. (in Russian).

KURBATOV, V.A., STARKOV, N.G.
[65] On the analytic representation of permutations, Uč. Zap. Sverdlovsk. Gos. Ped. Inst., 31, 1965, 151-158. MR 35 # 6652. (in Russian).

LANDAU, E.
[09] Handbuch der Lehre über die Verteilung der Primzahlen, Teubner 1909; reprinted by Chelsea 1953. MR 16 p.904

LAUSCH, H., MÜLLER, W.B., NÖBAUER, W.
[73] Über die Struktur einer durch Dicksonpolynome dargestellten Permutationsgruppe des Restklassenringes modulo n, J.reine angew. Math., 261, 1973, 88-99. MR 48 # 2231

LAUSCH, H., NÖBAUER, W.
[73] Algebra of polynomials, Amsterdam 1973. MR 50 # 2037.

LIDL. R.
[71] Über Permutationspolynome in mehreren Unbestimmten, Monatsh. Math., 75, 1971, 432-440. MR 46 # 5290.

[72] Über die Darstellung der Permutationen durch Polynome, Abhandl. Math. Sem. Hamburg, 37, 1972, 108-111. MR 46 # 9012.

[73] Tchebyscheffpolynome und die dadurch dargestellten Gruppen, Monatsh. Math., 77, 1973, 132-147. MR 47 # 6655

LIDL, R., MÜLLER, W.B.
[76] Über die Permutationsgruppen die durch Tschebyscheff-Polynome erzeugt werden, Acta Arith., 30, 1976, 19-25. MR 54 # 5196.

LIDL, R., NIEDERREITER, H.
[72] On orthogonal systems and permutation polynomials in several variables, Acta Arith., 22, 1972-3, 257-265. MR 47 # 6661

LIDL, R., WELLS, C.
[72] Chebyshev polynomials in several variables, J. reine angew. Math., 255, 1972, 104-111. MR 46 # 5291.

LUCAS, E.
[78] Théorie des fonctions numèriques simplement pèriodiques, American J. Math., 1, 1878, 184-240.

MACCLUER, C.R.
[66] On a conjecture of Davenport and Lewis concerning exceptional polynomials, Acta Arith., 12, 1967, 289-299. MR 34 # 7453.

McLEAN, D.W.
[80] Residue classes of the partition function, Math. Comp., 34, 1980, 313-317.

MAMANGAKIS, S.E.
[61] Remarks on the Fibonacci sequence, American Math. Monthly, 68, 1961, 648-649. MR 24 # A73

MANGOLDT, H. v.
[98] Dêmonstration de l'èquation $\sum_{k=1}^{\infty} \frac{\mu(k)}{k} = 0$, Ann. Scient. Éc. Norm. Sup., (3), 15, 1898, 431-454.

MATHIEU, E.
[61] Mêmoire sur l'etude des fonctions de plusieurs quantités, sur la manière de les former et sur les substitutions qui les laissent invariables, J. de Math., 6, 1861, 241-323.

MEIJER, H.G.
[70] On uniform distribution of integers and uniform distribution (mod 1)., Nieuw Arch. Wisk., 18, 1970, 271-278. MR 44 # 2712

MEIJER, H.G., SATTLER, R.
[72] On uniform distribution of integers and uniform distribution Mod. 1, ibidem 20, 1972, 146-151. MR 47 # 151

NARKIEWICZ, W.
[66] On distribution of values of multiplicative functions in residue classes, Acta Arith., 12, 1966/7, 269-279. MR 35 # 156.

[74] Elementary and analytic theory of algebraic numbers, Warszawa 1974. MR 50 # 268.

[77] Values of integer-valued multiplicative functions in residue classes, Acta Arith., 32, 1977, 179-182. MR 55 # 7956

[81] Euler's function and the sum of divisors, J. reine angew. Math., 323, 1981, 200-212. MR 82g # 10077

[82] On a kind of uniform distribution for systems of multiplicative functions, Litovsk. Mat. Sb., 22, 1982, 127-137.

[83a] Number Theory, Singapore 1983.

[83b] Distribution of coefficients of Eisenstein series in residue classes, Acta Arith., 43, 1983, 83-92.

NARKIEWICZ, W., RAYNER, F.
[82] Distribution of values of $\sigma_2(n)$ in residue classes, Monatsh. Math., 94, 1982, 133-141.

NARKIEWICZ, W., ŚLIWA, J.
[76] On a kind of uniform distribution of values of multiplicative functions in residue classes, Acta Arith., 31, 1976, 291-294. MR 58 # # 559.

NATHANSON, M.B.
[75] Linear recurrences and uniform distribution, Proc. Amer. Math. Soc., 48, 1975, 289-291. MR 51 # 379

[77] Asymptotic distribution and asymptotic independence of sequences of integers, Acta Math. Acad. Sci. Hung., 29, 1977, 207-218. MR 56 # # 11943

NIEDERREITER, H.
[70] Permutation polynomials in several variables over finite fields, Proc. Japan Acad., 46, 1970, 1001-1005. MR 44 # 5298

[72a] Distribution of Fibonacci numbers mod $5^k$, Fibonacci Quart., 10, 1972, 373-374. MR 47 # 3303

[72b] Permutation polynomials in several variables, Acta Sci. Math. (Szeged), 33, 1972, 53-58. MR 46 # 8998

[75] Rearrangement theorems for sequences, Astêrisque 24-25, 1975, 243-261. MR 52 # 8068

[80] Verteilung von Resten rekursiver Folgen, Archiv f. Math., 34, 1980, 526-533. MR 82c # 94011

NIEDERREITER, H., LO, S.K.
[79] Permutation polynomials over rings of algebraic integers, Abh. Math. Sem. Hamburg, 49, 1979. 126-139. MR 80k # 12002

NIEDERREITER, H., ROBINSON, K.H.
[82] Complete mappings of finite fields, J. Austral. Math. Soc., 33, 1982, 197-212.

NIEDERREITER, H., SHIUE, J.S.
[77] Equidistribution of linear recurring sequences in finite fields, Indag. Math., 39, 1977, 397-405. MR 57 # 3085.

[80] Equidistribution of linear recurring sequences in finite fields, II, Acta Arith., 38, 1980, 197-207. MR 82f # 10048

NIVEN, I.,
[61] Uniform distribution of sequences of integers, Trans. Amer. Math. Soc., 98, 1961, 52-61. MR 22 # 10971

NÖBAUER, W.
[64] Zur Theorie der Polynomtransformationen und Permutationspolynome, Math. Ann., 157, 1964, 332-342.

[65] Über Permutationspolynome und Permutationsfunktionen für Primzahlpotenzen, Monatsh. Math., 69, 1965, 230-238. MR 31 # 4754

[66] Polynome, welche für gegebene Zahlen Permutationspolynome sind, Acta Arith., 11, 1966, 437-442. MR 34 # 2562

PILLAI, S.S.
[40] Generalization of a theorem of Mangoldt, Proc. Indian Math. Soc., sect.A, 11, 1940, 13-20. MR 1 p.293

RAYNER, F.
[83] Weak uniform distribution for divisor functions, to appear

RIEGER, G.J.
[77] Bemerkungen über gewisse nichtlineare Kongruenzen, Elem. Math., 32, 1977, 113-115. MR 57 # 3054

[79] Über Lipschitz-Folgen, Math. Scand., 45, 1979, 168-176. MR 82a # 10055

ROBINSON, D.W.
[66] A note on linear recurrent sequences modulo m, American Math. Monthly, 73, 1966, 619-621. MR 34 # 1260

ROGERS, G.L.
[91] Messenger of Math., 21, 1891-1892, 44-47.

SATHE, L.G.
[45] On a congruence property of the divisor function, Amer. J. Math., 67, 1945, 397-406. MR 7 p.49

SCHMIDT, W.
[76] Equations over finite fields: an elementary approach, Lecture Notes in Math. 536, 1976. MR 55 # 2744

SCHUR, I.
[23] Über den Zusammenhang zwischen einem Problem der Zahlentheorie und einem Satz über algebraische Funktionen, Sitz. Ber. Preuss. Akad. Wiss. Berlin 1923, 123-134 = Ges. Abh., II, 428-439, Springer 1973.

[33] Zur Theorie der einfach transitive Permutationsgruppen, ibidem 1933, 598-623 = Ges. Abh. III, 266-291, Springer 1973.

[73] Arithmetisches über die Tschebyscheffschen Polynome, Ges. Abh. III, 422-453, Springer 1973.

SCOURFIELD, E.J.
[74] On polynomial-like functions weakly uniformly distributed (mod N), J. London Math. Soc., 9, 1974, 245-260. MR 50 # 7001

SERRE, J.P.
[68] Une interprétation des congruences relatives à la fonction $\tau$ de Ramanujan, Séminaire Delange-Pisot-Poitou 9, 1967/68, nr.14, 1-17. MR 39 # 5464

[72] Congruences et formes modulaires, Sém. Bourbaki, 24, 1971/72, nr.416. MR 57 # 5904

[75] Divisibilité de certaines fonctions arithmétiques, Sém. Delange--Pisot-Poitou: 1974/75, Théorie des Nombres, Exp.20, 28pp.

SHAH, A.P.
[68] Fibonacci sequence modulo m, Fibonacci Quart., 6, 1968, 139--141. MR 40 # 86

ŚLIWA, J.
[73] On distribution of values of $\sigma(n)$ in residue classes, Colloq. Math. 27, 1973, 283-291, corr.332. MR 48 # 6044

SWINNERTON-DYER, H.P.F.
[73] On $\ell$-adic representations and congruences for coefficients of modular forms, In "Modular Functions" III, Lecture Notes in Math., 350, 1973, 3-55. MR 53 # 10717

[77] On $\ell$-adic representations and congruences for coefficients of modular forms, II, In "Modular Functions V", Lecture Notes in Math., 601, 1977, 63-90. MR 58 # 16520

UCHIYAMA, M., UCHIYAMA, S.
[62] A characterization of uniformly distributed sequences of integers, J. Fac. Sci. Hokkaido, 16, 1962, 238-248. MR 27 # 1433

UCHIYAMA, S.
[61] On the uniform distribution of sequences of integers, Proc. Japan Acad., 37, 1961, 605-609. MR 25 # 1145

[68] A note on the uniform distribution of sequences of integers, J. Fac. Sci. Shinshu Univ., 3, 1968, 163-169. MR 40 # 115

VAN DEN EYNDEN, C.L.
[62] The uniform distribution of sequences, Diss., Univ. of Oregon 1962.

VEECH, W.A.
[71] Well distributed sequences of integers, Trans. Amer. Math. Soc., 161, 1971, 63-70. MR 44 # 2715.

VINCE, A.
[81] Period of a linear recurrence, Acta Arith., 39, 1981, 303-311.

WALL, D.D.
[60] Fibonacci series modulo m, American Math. Monthly, 67, 1960, 525-532. MR 22 # 10945

WARD, M.
[31a] The characteristic number of a sequence of integers, satisfying a linear recursion relation, Trans. Amer. Math. Soc., 33, 1931, 153-165.

[31b] The distribution of residues in sequences satisfying a linear recurrence relation, ibidem, 33, 1931, 166-190.

[33] The arithmetic theory of linear recurring series, ibidem, 35, 1933, 600-628,

WEBB, W.A., LONG, C.T.
[75] Distribution modulo p of the general linear second order recurrence, Atti Accad. Lincei, 58, 1975, 92-100. MR 54 # 7396

WEGNER, U.
[28] Über die ganzzahligen Polynome, die für unendlich viele Primzahlmoduln Permutationen liefern, Diss. Berlin 1928.

WEIL, A.,
[48] Sur les courbes algébriques et les variétés qui s'en deduisent, Actual. Sci. Industr. v. 1048, 1948. MR 10 p.262

WELLS, C.
[67] Groups of permutation polynomials, Monatsh. Math., 71, 1967, 148-262. MR 35 # 5421.

[68] Generators for groups of permutation polynomials over finite fields, Acta Sci. Math. (Szeged), 29, 1968, 167-176. MR 38 # 5903.

WILLETT, M.
[76] On a theorem of Kronecker, Fibonacci Quart., 14, 1976, 27-29. MR 53 # 264

WILLIAMS, K.S.
[68] On exceptional polynomials, Can. Math. Bull., 11, 1968, 279--282. MR 38 # 140

WIRSING, E.,
[67] Das asymptotische Verhalten von Summen über multiplikativen Funktionen, II, Acta Math. Acad. Sci. Hung., 18, 1967, 411-467. MR 36 #6366.

ZAME, A.
[72] On a problem of Narkiewicz concerning uniform distribution of sequences of integers, Colloq. Math., 24, 1972, 271-273. MR 46 # 5272.

# INDEX

A D D E N D A

1. p.2. It was recently shown by I.Ruzsa (On the uniform and almost uniform distribution of $(a_n x)$ mod 1, Sém.Théorie des Nombres,1982-3,exp.20,Université de Bordeaux 1983.) that there exists a sequence $a_n$ of integers which is UD(mod N) for all N whereas $(a_n x)$ is UD(mod 1) for no real x.

2. p.3. A study of uniform distribution of sequences of algebraic integers was carried out in H.NIEDERREITER,S.K.LO, Uniform distribution of sequences of algebraic integers, Math.J.Okayama Univ.,18,1975,13-29.

3. p.10. Cf. also H.NIEDERREITER, On a class of sequences of lattice points, J.Number Theory 4,1972,477-502.

4. p.26, line +6. H.NIEDERREITER [70] should be also quoted here.

5. p.32. M.Hall's result was generalized by H.NIEDERREITER (On the cycle structure of linear recurring sequences, Math. Scand.,38,1976,53-77).

6. Problem I, stated on p.9 has just been solved independently by E.Rosochowicz and I.Ruzsa.